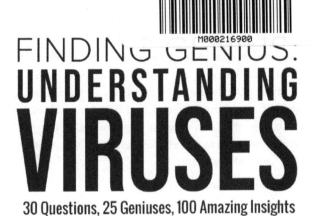

FINDING GENIUS:
UNDERSTANDING
VIRUSES

30 Questions, 25 Geniuses, 100 Amazing Insights

by Richard Jacobs, et al.
edited by: Lindsay Hoeschen

Finding Genius Foundation
8530 Barasinga Trail
Austin, TX 78749
(888) 988-7381
www.findinggeniusfoundation.org

Publisher: Finding Genius Foundation,
a 501(c)3 Nonprofit (pending approval by IRS)

Written by: Richard Jacobs, Host: Finding Genius Podcast;
Director: Finding Genius Foundation
Co-Authored by: Denis Noble, Shiraz Shah, Adolfo Garcia-Sastre, Joseph R. Masci, St Patrick Reid, Jeremy Barr, Lily Wang, Eugene V. Koonin, Nathalie Gontier, Paul Offit, Marilyn J Roossinck, Gareth Brady, Luis Villarreal, Michael Betts, William B. Miller, Nils G. Walter, Guenther Witzany, Forest Rohwer, Robert Siegel, Paul Turner, Dr. Richard Allen White III, Seyedtaghi Takyar, Matthew B. Frieman, James Shapiro

Edited by: Lindsay Hoeschen

Audiobook version by: Brian J. Ford, FLS HonFRMS

Ordering Information:
Quantity sales. Special discounts are available on quantity purchases by corporations, associations, and others. For details, contact the publisher at the address above.

Orders by U.S. trade bookstores and wholesalers. Please call (888) 988-7381 or visit www.findinggeniusfoundation.org.

Printed in the United States of America

Published in 2021

ISBN: 978-1-954506-03-9

DISCLAIMER

This book is provided for informational purposes only. Do not rely upon the information contained within it to make a medical, health-based, financial, lifestyle, or any other decision. Consult with licensed professionals regarding any health questions or issues you may have. Reading this book does not constitute the practice of medicine, and no doctor-patient relationship is implied nor formed by reading this book.

Many, most, or perhaps all of the co-authors of this book work within the University system, consult with various corporations, may seek government or privately-funded grants, and must be extremely careful about their opinions, thoughts, observations, and answers to the questions in this book.

Their words in this book may conflict with the views of some or all of the organizations with which they are involved; readers should not take anything written in this book to be the firm belief of any of the co-authors.

Organizations involved with these co-authors should also understand that the views expressed in this book should be held completely separate from their consideration of their relationships with the co-authors.

All co-authors have been asked to speculate and "let their hair down" in answering these questions to the best of their abilities. They do not claim to be experts based on the questions asked, and specifically disclaim the providing of any advice – this is pure information and speculation.

DEDICATION

Most folks I've interviewed would blush or chafe against me calling them a genius, and that's understandable. Calling yourself a genius is narcissistic; but when others say you're a genius, beyond a slight embarrassment, you are aggrandized in the eyes of others.

To my co-authors: I say it on your behalf – "You are all geniuses in what you do!" You, the reader, will likely form similar opinions as you read through this work.

My 15-year hobby, including 4 years of 2,700+ podcasts (similar to Tim Ferriss' tropism), is to find, interview, learn from, and disseminate the amazing knowledge I've gained from people I consider to be geniuses in their field(s) of inquiry.

As mentioned in the disclaimer page, fellow co-authors, I take your participation in this book very seriously. The last thing I want to do is represent you incorrectly, or cause friction in your academic, professional, or personal life.

Readers will notice that no single question contains all co-author answers; please do not take this to mean that and individual's answer wasn't a great one. For the sake of time, brevity, flow, and other issues, I could not include everyone's answer to every question.

Thank you for participating in this book as co-authors, especially with an 'outsider' to your field. Thank you as well for your bravery in speculating and providing your best answers, spur-of-the-moment, under the pressure of an interview situation, to answer as best you could.

I tried my best to make the questions relevant now, and for the next 20 years at least. I believe this book will NOT be substantially outdated in 1 year, but perhaps in 20+ years.

My goal is that your participation in this book will advance your reach and amplify your voice, and, above all, help inform curious amateurs and science professionals about the amazing world of viruses.

This project felt like the Olympics of Interviewing. It was thrilling (in the nerdiest sense of the word) to hear such divergent and intelligent answers to super-difficult and speculative questions. Thank you for your participation, for all that you do, and for making this experience a big-time thrill as well as a deep dive into knowledge of viruses.

TABLE OF CONTENTS

7

BONUS CONTENT

To listen to the all of the full audio interviews and transcripts that comprise this book (and more), visit: FindingGeniusFoundation.org and select "Publications" from the menu.

INTRODUCTION

 Five years ago, I didn't even know the difference between viruses and bacteria. Since that time, through my interviews with over 2,700 researchers, clinicians, CEOs, CSOs, doctors both allopathic and naturopathic, scientists, and well-studied amateurs afflicted with various health conditions, I've acquired a "street MBA" or "street PhD" in various biosciences.

...with no formal credentials beyond a B.S. in Chemical Engineering and a street PhD, it would be foolish for me to write a book on Understanding Viruses without the help of geniuses in the fields of Evolutionary Biology, Vaccines, Bioinformatics, and of course, the viruses of plants, bacteria (phage), fungi, viruses of viruses, and viruses of animals and humans.

However... one of my very special, Liam-Neeson-type skills, is to ask questions (of scientists with decades of experience), that make them exclaim: "Huh... That's a VERY good question!"

This book is the result of 30+ interviews of genius-level science professionals in various fields. They have all been asked the same set of questions, and have answered, to the best of their ability and knowledge.

What you're about to read is the answers to these questions, not from all the individuals, but from the most interesting 4 or 5 who answered each question.

The questions do NOT include all interviewees' answers, for several reasons:

- The best scientists know which areas they are experts in, and which to defer to others
- Some answers are similar and can be combined
- 4-5 answers per question makes this book more readable, yet plenty long and informative enough without being exhausting

Having interviewed 2,700+ guests over the past 4 years, I've noticed various trends and curiosities. One of my favorite questions to ask guests is: "What do you notice about your field of study / research / clinical work that surprises you?"

<u>Here are some eyebrow-arching insights:</u>

When first interviewing individuals in a field, I may know nothing, and resort to asking basic questions. At this point, luck and several gracious, accommodating, well-spoken guests help me to start asking better questions of future guests. Once I reach 15 interviews in a subject, it becomes familiar.

At 30 interviews, a picture forms of the 'ant farm' of activity in a given field – who is tunneling here, who is building there, and what all the creatures are working on. 30+ interviews in, guests start to say my favorite response: "Huh... That's a good question!" and I know I'm providing new questions for my experts to ponder (it's my way of 'giving back', and of course, an ego-stroke as well).

In some fields, after interviewing 100+ people, I start to sound like an insider, and I've run into 5 to 6 super-knowledgeable guests who are also well spoken and articulate; these are <u>the geniuses of their field</u>.

When I speak to a genius, I can hear it and feel it within the first 5 minutes of the interview. Interviewing these people is thrilling. When I ask a question that 20 previous interviewees failed to answer, the genius says, "Oh yes! I've observed that, and here's an example."

Somehow, the genius answers skew towards the positive – towards the "Yes, of course, and here's where, when and how it happens." – the non-geniuses tend to give far more negative answers. To me, that's a sign of a closed mind (not in all cases of course).

How to Become a Scientific Genius

Scientists need to interact with and talk to other scientists far more than they currently do. If I was an expert in Oncology, and I answered interview questions with, "I don't know", "That's not my area", "I'm not an expert in X", or "I don't think so", while another Oncologist has the answers to these questions, it means I haven't spent enough time talking with others in my field to learn beyond my narrow scope of research or clinical knowledge.

You would be amazed how often I get <u>this exact pattern of response</u>: 5 experts tell me: "I don't know", while the 6th expert says, "Yes, of course that happens. Here are the details". Get out there and talk to your colleagues and people in ancillary fields!

Reading papers in journals is just the start, and pales in comparison to talking to other experts. Email the authors; get them on the phone; ask questions that come from reading their paper. There is a world of information and depth of knowledge missed by only reading journal articles.

If you're truly curious about a scientific paper, article, or topic, talk to that person (their time permitting), and dig deep on their assumptions, their results, their missteps, and their new view of that subject. It will transform a 2-dimensionsal, black and white view to a 3D, full color view of the subject.

If you're not an expert in a field that is closely linked to your own, start learning about it. Why? Because the more you learn that is outside your immediate scope of research or clinical work, the more associations you can make, common threads you can find amongst all science, and the more likely it is that you will solve your research, clinical, or commercial problems vs. other people who stay in their lane.

Why should it take a massive, societal earthquake like the 2020 pandemic to bring scientists together, or to get people interested in science?

I strongly believe that it's everyone's responsibility who loves science, to promote it for the benefit of humankind. Communication, collaboration, coordination, and interaction of all sorts is the accelerates scientific progress, which can benefit humanity.

Imagine if you and your peers (or other like-minded, science-curious friends) spent focused time together, twice a year, virtually or in person, to discuss each other's work and provide insights? Scientific progress would rocket forwards, I believe.

15

What does it take to learn and start to master a subject?

All it takes to learn science is to be interested in and curious about a subject, then to ask questions about that subject, to read, to listen, to learn, and even to experiment where you can.

A favorite quote of mine comes from Confucius, who said: *"The man who asks a question is a fool for a minute, the man who does not ask is a fool for life."*

...this means that, even if you know nothing about a subject, don't be afraid to ask. Of course, you should respect the time of experts and do your homework first – read their materials, even if you understand only 10%, watch videos about them, use their products if they have created them.

Once you have done your homework, you'd be surprised at how often experts are thrilled for someone to have genuine curiosity about their field of expertise. Many an interviewee has told me that there are few people (sometimes no one) who are interested in talking with them about their work.

Not all experts will be nice to you. Some will be dismissive, yet others will be welcoming. Be the fool for a minute instead of the fool for life – ask questions!

What level of sophistication do you need to read this book?

Finding Genius Podcast interviews are for two kinds of audiences: 1) curious amateurs, and 2) science, engineering, or other professionals who have a general knowledge of and curiosity about science, but who are not experts in the field being covered.

At the start of my journey in science, I already had a B.S. in Chemical Engineering, but that was from 1998, certainly not 2016 when I started. Reading scientific papers was difficult at first – I understood about 10% of what I was reading. As I interviewed more scientists, researchers, clinicians and others, I started to pick up the vocabulary and thinking process needed to read scientific papers.

After a year of reading, my comprehension increased, and now stands at around 70%, typically. In the areas of Physics or other sciences in which I'm not conversant, I understand about 30%. In advanced math, since it's a language unto itself, especially at high levels, I am back to the 10% level.

Apply yourself, and you'll be surprised at what you can learn in a year's time.

Orthodoxy, Assumptions, and Anthropomorphic Thinking:

I have the sneaking suspicion that young scientists, during their formative schooling years, are taken into basements, tied up, and beaten until they agree to abide by the following dogma:

- **Anthropomorphic thinking is bad** – bacteria, our cells, viruses, fungi, and most other forms of life are inferior to humans and we should never ascribe human-like qualities to anyone or anything except human beings.

17

- **Biology is stochastic / random** – biological processes are no different than non-living systems. All operate based on the laws of physics, thermodynamics, and other sciences. Nothing is deterministic, deliberate, based in thought, cognition, agency, or other human-like characteristics. Life is akin to a sophisticated machine – it has no "spark", no "soul", no anthropomorphic qualities.
- **Natural selection over time** – Natural selection governs evolutionary development and explains much of human behavior. Evolution happens on enormous time scales of millions of years and is driven by random mutations acted upon by natural selection.
- **Speculation is bad** – stay in your lane, stick to your field, leave everything else to qualified experts. They should behave the same way when it comes to interacting with you. Speculating is not 'good or proper science', and everything is anecdotal (pejorative) unless a double blind, placebo-controlled clinical trial is run to establish a hypothesis as a theory.
- **Science and religion are incompatible** – theological considerations have no place in science.

I'm NOT telling you to abandon these thoughts. I AM telling you to <u>include both these thoughts, and their opposites</u>, when reading this book, and in your work and thoughts surrounding science.

I'm not remotely omniscient, and I certainly don't know even a tiny smidgen of everything out there, but what I do know is this: The more open-minded you are about science, the more possibilities will arise in your thinking, the more open-minded and creative you will be in finding solutions to your problems; you will be more likely to seek outside counsel from other experts, and people in vastly different fields.

Innovation often comes from people outside an industry. One of the reasons is that those individuals aren't chained to the orthodoxy, assumptions, and thinking that govern a single industry.

This may come across as arrogant, but it needs to be said... a significant percentage of the scientists / researchers / clinicians / CEOs and others I've interviewed are going to be charging down dead end alleys for years to come, because they are too provincial in their thinking and assumptions.

This is exactly how I was able to come up with the questions for this book – by not adhering to doctrine, by being open-minded, deliberately anthropomorphic in my thinking, and being willing to ask super-tough questions.

How do I know that I am on the right track? Because every single one of the interviewees in this book told me, during the interview, at various times and in various ways:

"Huh... that's a very good question."

"I don't know – I need to think about that."

"I don't even know if this question is answerable!"

"You ask some great questions."

"Now you're getting outside my area of expertise."

Many of the interviewees also LAUGHED at some of my questions. They didn't laugh to make fun of me – they laughed as if to say: "Damn... You got me."

So please, read this book with an open mind. As a magician might say: "suspend your disbelief".

Last Note Before We Begin:

I'm going to weigh in at the end of each question and give my opinion. How dare I? I was hesitant to do so, but my friend and mentor, Bill Miller suggested that my speculation would open the door for other people to speculate on the answers to these questions, too.

...and THAT is what I want. Keep reading. Speculate. Think. Take these answers back to your own lives, your own research, your own work. Use them to fuel more scientific discovery. Obsolete this book as fast as you can, for the good of humanity.

QUESTION 1:

"Are viruses alive?"

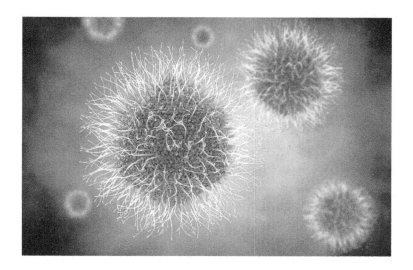

Richard Jacobs:

I'm weighing in on this question before hearing the answers from other scientists. For all other questions in the book, I will wait until my co-authors answer before sharing my thoughts. Why insist on emphasizing this question in particular?

The specific issue of whether viruses should be considered to be living or not motivated this book and all the questions that I have asked the experts to consider. Without this level of open-mindedness in viewing viruses as living entities, there probably wouldn't be a book at all.

There are three ways of looking at viruses: they are either altogether alive, they are altogether not, or they are contingently alive in cells when connected to cellular resources. When a scientist asserts that viruses are not alive, explanations of their effectiveness in biology have to rely on random variations, rather than by self-directed actions.

Their interaction with biological systems would depend on the law of large numbers, and in the case of viruses, this is a huge factor because they appear to be present on Earth in the concentration of 1023 to 1031 — an unbelievably large number of viruses.[1]

They also must rely on natural selection. If viruses are not alive and have no agency, no thought, and no goals, then only natural selection will put pressure on them and cause them to change over evolutionary time. Viruses actually change extremely quickly, but under this view, natural selection must be responsible for it.

There is nothing wrong with the answers or ways of thinking that are pinned by a belief that viruses are not alive. However, I've found that most of the people who answer in the negative to this question are unable to answer many of the other questions in this book, or will give conventional Neo-Darwinian answers such as, "It's just natural selection," or "It's just random."

I choose to believe that viruses are alive. Here's why I think it is important and deserves first scrutiny before all that follows. It frees our minds to openly consider both sides of this vital question.

I can certainly incorporate everything into the "No" category; I can incorporate random events, large numbers, natural selection, and selection pressure into an evolutionary narrative but I can also raise substantial other questions in this book and attempt to have them answered with similar freedom of thought. In that regard, I have received insightful responses from all the scientists in this book.

Why aren't viruses considered to be alive? Some people say, "What is your definition of 'alive'?" or "What is your definition of 'life'?" That's a good question. Some people say it is the ability to reproduce independently, and it appears that viruses cannot reproduce without infecting a cell, whether that's a bacterium, another virus, a human cell, an animal cell, a plant cell, fungi, etc. That's one reason people say that viruses are not alive.

Another group of people say they are not alive because they have no cognition and no agency but are mere chemicals that follow the laws of physics and are subject to evolution and the central dogma, just like all other cells and living things.

I was sitting in my car one day and I looked at the trees across a field from me; it wasn't a windy day, and I observed no movement in the trees, therefore I could have said that a tree is not alive, despite knowing that it is completely alive.

If I had zoomed in and gotten really close to the tree, I might have seen movement of the leaves and limbs of the tree, and I might have said that the tree is alive because it moves, albeit slightly. If I were to zoom in to the microscopic level, I would see all kinds of bacteria and cells, and would be able to say that the tree is very much alive.

Then I thought to myself, what if the tree grows a fruit and has seeds? If I did the same experiment and zoomed in on a seed, I may not see anything move, and therefore I'd think the seed is not alive. But as we all know, seeds are a seemingly inert transitory state of the living form of plants and trees. I have thought about viruses and considered whether viruses either outside of infected cells, might themselves be another living entity even though they appear to be inert.

If I thought of viruses mostly in this state and thought it rare that they enter into a cell, activate, and replicate, then I might superficially assess that they are not alive, and it would probably make sense to have that perception. What if I saw trees mostly in the seed state, and only very rarely, for a short period of time, saw them in the fully grown or living state?

I might think that plants are not alive, because for the most part, they are just inactive, non-motile seeds; only every once in a while, under the right conditions (e.g. soil, water, sunlight) — just like a virus inside of a cell — they become alive. I would say that perhaps trees and plants are contingently alive.

This argument can be applied to other animals. For example, if a bear is hibernating for six months, its heart rate and temperature drop, and it may be almost absolutely still. Would you consider the bear to be alive if you were unaware of its biological movements and history and were just observing it de novo within this one limited state? Perhaps, there are states that may be living but don't meet our uncertain criteria.

I think that we should consider viruses, both inside and outside of cells, as living things. Perhaps viruses are not alive when they are in the virion stage but are contingently alive only when they enter into a cell, where they have the ability to replicate and use the cellular machinery like a parasite. This type of thinking can lead to many useful ideas.

Let us now return to the principle of 'anthropomorphic thinking'. If I assume that viruses are alive and in their own special way have an analog of all the faculties and capabilities that animals or bacteria have, then a lot of ideas come to mind.

For instance, I can think that they know 'self' from 'other,' and we may see this with regard to viral quasispecies, and their ability to hide from the immune system. Hiding from the immune system or putting the immune system to sleep is a way the virus defends itself, and in my opinion, you have to contemplate a 'self' if you are going to do that.

Viruses also have the ability to — as my friend and mentor Bill Miller likes to say — "evaluate ambiguous information," which definitely appears to be the case when considering their ability to go from latent to virulent or pathogenic in the presence of certain cues.

Viruses possibly have memory, because they are not just one kind of virus with one nucleotide sequence but appear to be a swarm of slightly or substantially different nucleotide sequences. Perhaps this allows them to more adequately defend against the immune systems of people, plants, or animals that they enter into, because all those creatures of a given species are different on a biological level.

Do viruses care for their progeny? I don't know. Some people say that all viruses want to do is replicate, but they are saying they want to replicate, and why would viruses want anything if they are not alive? And if they do want to replicate, they'd want to make good copies of themselves, not non-viable, mutated forms incapable of infecting other cells.

They may not be a good 'parent' virus, but all the same they may be orchestrating the creation of progeny in a way that is adaptive to the conditions that they find themselves in, and in turn, this makes it more likely that their progeny will survive and spread. This could definitely be considered caring for their progeny. I believe viruses do have a sense of self, even if it's at the swarm or community level instead of the individual level.

Viruses appear to take coordinated action.[2] We don't know if there is viral quorum sensing like there is in bacteria, but there is a latency period between the time you are infected by a virus and the time you actually get sick. This period of time varies, and we don't know why. It could be that there has to be a lot of viral replication, or it could be coordinated action, whereby the viruses are signaling each other within a given cell, across cells, etc.

Viruses appear to have knowledge of the outside world. The tropism of some viruses is matched to the particular types of cells that it infects.[3] For example, flu virus may infect respiratory cells, but it also has a tropism for those kind of cells. As another example, rabies may cause its victims to bite other victims, and has a tropism for the salivary glands and cells involved in that behavior.

In order to be successful, viruses want to spread from host to host, and especially when it comes to zoonotic events, viruses appear to have knowledge of the outside world. Parasites appear to also, which is how they are able to go from host to host and back again in a cycle, and thrive in completely different environments.

There is a seed-and-soil model that applies to cancer metastasis, with the idea being that in the right soil, the right seed will grow. Viruses—just like trees that need soil and sunlight and nutrients—need the environment of a cell in order to build more of themselves, perhaps do sensing, and to live and thrive.

I thought the 'one virus, one cell infection' model was the only right model, but a number of scientists have told me that multiple viruses can infect a cell. In some cases, a virus will even defend the cell against other viruses,[4] like a dog guarding a bone.

Again, this shows a sense of 'self' or 'other.' In plants, successful infection sometimes depends on multiple types of viruses infecting a cell, with each one carrying a part of the genetic payload needed to create new progeny virions.

There definitely may be coordinated infection, perhaps at the cell membrane level or the cell wall level. Perhaps multiple viruses are needed to coordinate entry; we just don't know. It seems like multiple viruses can enter and exclude from entry viruses which are different from them.

Perhaps viruses in their virion stage are like sperm, where a whole swarm of them look for target "egg" cells, and then fuse with and enter that cell to produce new progeny. In this way, the sperm-egg-zygote model could be an analog to the virus-cell interaction model, where the virus inside a cell is considered a vigote instead of a zygote (a term I just coined).

When I consider viruses to be alive, I'm able to fully consider the possibility that they may be capable of extracellular sensing (i.e. sensing outside the cell) and reconnaissance. Perhaps viruses can harness and alter the contents of the extracellular vesicles that cells put out. We know that bacteria transmit plasmids, and that human and animal cells put out extracellular vesicles of all kinds.

Perhaps viruses, by altering the central machinery of the cell, can create customized extracellular vesicles to communicate with other viruses and coordinate action, coordinate infection, coordinate latency, coordinate virulence, etc.

My bias is to consider viruses as alive. It is an open stance that energizes all the further questions on the next pages. I encourage you, the reader, to make your own determination by reading the coming answers to this question. I believe that this is the foundational question of the entire book, and there are many more questions to follow if you are willing to be open to the fact that viruses are indeed alive.

Luis Villarreal: (bio page 203)

The capacity for persistent inapparent viruses to cause harm in infections of others never goes away; it's an inherent selective feature that they maintain. For example, when it comes to mouse hepatitis virus (a coronavirus) in a population of mice that is persistently affected and showing no disease, the virus retains the capacity for causing disease in other populations. It will not lose this disease capacity with continued persistence and continued passage, and I don't think that's going to happen with COVID-19 either.

These viruses are unendingly dynamic in terms of the composition of their RNA. They are never clones. Although there is one consensus type, which can be rather stable and is the most abundant version of the sequence, you always have lots of variants. This is true for Covid-19. As soon as you passage the virus, you generate a quasispecies that has a lot of minorities and variants. Most of these variants don't have any phenotypes that we are aware of, but they are interacting with the population; sometimes they interfere and sometimes they complement, but they do all kinds of things within the context of that population.

There is evidence that the quasispecies is dynamic as it moves through tissue in one individual infection. For example, mouse hepatitis virus has the capacity to come in through the gut and move through nervous tissue. These experiments were done with poliovirus in order to test the importance of this quasispecies dynamic in the emergence of disease.

Since poliovirus has been studied a lot, it's possible to make a version of the polymerase that has high fidelity and a much lower error rate. Then, you can turn it back into a virus and use it to infect a mouse and study the resulting pathology.

With polio, the pathology that is particularly interesting is the neuropathology in which infection spans the spinal column and gets into the central nervous system. It was observed that if there was a high-fidelity polymerase, the virus was unable to truly replicate far beyond the initial site of infection. It did not seem to be adaptable to the neurological infection, and the disease that would follow.

William B. Miller, Jr.: (bio page 206)

I believe that viruses are alive within their cellular context. Why might I defend that? The crucial attributes of life are the ability to communicate, to reproduce, and to problem-solve. Viruses seem to be capable of each of those things. They can make contingent decisions within the cell.

Can they do that outside a cell as a free-living virion or within an extracellular vesicle? I don't know that answer. That's certainly something that requires further research. However, it is certainly more productive to consider viruses as being alive (in the sense of a basal purposefulness) within a cell since this better explains the range of reciprocal reactions between cells and viruses.

Within the cell, the virus becomes a kind of co-participant in cellular life (cells are clearly alive and no one is going to argue that they are not) and they share the metabolic apparatus of with the cell. Each co-respondent both the virus and the cell has co-participatory status. This is why I think they are alive within the cell.

There is an excellent concept by a French researcher named Patrick Forterre, who has championed the virocell concept. Basically, Forterre views viruses as being in a co-relationship with an infected cell. This is a dominant form of co-dependence that exists across all of the cellular domains. The virus and infected cell form a novel type of hybrid cellular organism, the virocell, which he regards as a living entity.[5]

Michael Betts: (bio page 223)

No. In my opinion, they are basically machines, and they do not have classical things like organelles that even the primitive cells would have. They do have membranes, but they take those from the cells they are budding from. They are never autonomous; they always require something from a host cell. I would not classify them as being alive whatsoever. They have components necessary for life, like DNA or RNA, but they are not alive.

I don't think I would consider a virus alive even when it's infecting a cell. I think it's a combination of proteins and genomic material that are interdependent and therefore they would have to be interdependent for the assembly process to take place inside of a cell. Each of these components has somehow evolved to interact with the others and form these complexes, but I don't consider them alive, even in that context.

Richard Allen White III: (bio page 209)

Without a doubt, they are alive. I believe in Patrick Forterre's work from 2010. He published a paper that definitely needs to be republished with some more modern examples, but the idea is that at the end of the day, it's important to include viruses in pantheon of life.

These things will continue to be there and they will be there long after we're gone, when our planet is engulfed by the sun and forms a red giant billions of years from now. I think we've been teaching virology wrong and Patrick's work has really shed light on this.

In general, it goes like this: if you see pollen on a car, would you call pollen trees? Absolutely not. If you see a basket of eggs from a chicken, would you call those eggs chicken? Absolutely not. This is what we've been doing with viruses; we've been saying the virion — the replication unit of the virus — is the virus. The virus is the infected cell.

For classic examples like polio, the virus enters, infects, and takes over the host machinery such that within 20 minutes after infection (maybe even an hour in polio), the cell is no longer making cellular proteins; it is making only viral proteins. It makes so much so that the cell expresses these viral proteins on the outside surface of the cell, and then it can actually go infect other cells. We end up with cell-to-cell infection. This is found in HIV in the mega cells called syncytium. When we have very high titers of virus in cell culture, we can actually see the phenomenon.[6]

Güenther Witzany: (bio page 210)

Definitely. I believe viruses are alive because they depend on a genetic code. No natural code codes itself and no natural language speaks itself; you always need a biotic agent which is competent to use this code. From my perspective, the genetic code has edited us, and viruses and their relatives, such as transposable elements, are the agents that edit the code of the host organisms.

Within viral communities, viruses compete naturally but cooperate to reach their goals, such as to infect. In many cases, they have helper viruses, so together they can invade a host organism. Viruses are not alone if they infect a host, because there are also competing viruses which want to infect the same host. They have to search interactional patterns to reach their goal. All of this shows clear biotic behavior in contrast to abiotic molecules, which cannot communicate.

Robert Siegel: (bio page 227)

I'll begin by saying that I think it is a false dichotomy to ask if a virus is alive or not alive. I will posit that viruses are on the hazy border between what's living and what's not living. I often teach with poetry, and this is a stanza from a poem I wrote called The Secret Life of Viruses:

- A virus has a structure giving it
- Properties that lie between:
- Inanimate and truly living -
- Less than cell but more than gene

Many people who try to answer this question don't define the term "living." In order to get a really effective answer to this question, we need to agree on what we are talking about. I have a number of different definitions. This question is interesting from a scientific standpoint because it allows us to understand the essential features of "living."

It is also essential from a pedagogic standpoint because it's useful in teaching about the nature of viruses. Of course, it is also interesting from this linguistic standpoint that I am talking about, namely, how do we define things like viruses? How do we define things like 'life'?

Let me pose three different definitions of "life". The first of these definitions was formulated in the mid-1800s by Schleiden and Schwann who proposed that "all living things are comprised of cells".[7] For the biological world, that appears to be true—for bacteria, plants, humans, and essentially everything that we can think of. But viruses are not comprised of cells; they are smaller than cells, and they have to enter cells and take them over in order to make more virus particles. By that first definition, viruses are not alive.

According to my second definition, we can define life by looking at "vital properties" which is to say, all the properties that living things have in common. These would include replication, and, of course, viruses are really good at that. But it would also include things like eating, excreting waste, sensing the environment, homeostasis (self-regulation), and movement, and viruses don't do any of those things. They don't metabolize energy, and they don't have any mechanism of synthesizing protein or other molecules. By the second definition, they fail miserably.

My third definition focuses in on this whole idea of reproduction or replication. We can ask: how it is that living things replicate? For this, it is useful to look at the paradigm of Francis Crick, who came up with an idea called the Central Dogma of Biology (which you know must be important with a name like the Central Dogma of Biology).

The Central Dogma of Biology

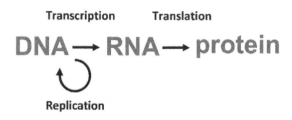

Basically, the Central Dogma describes the flow of biological information. It turns out that in all living things, the biological information that is passed on to the next generation is stored in the form of DNA, and that information is then read out, like a Xerox copy, in the form of messenger RNA (mRNA). The instructions in messenger RNA are then used to make protein.

Of course, in order to be able to pass the information on, the DNA has to be able to replicate. To summarize, all living things have their information in the form of DNA. They all carry out the three processes of DNA replication, expression through mRNA, and synthesis of proteins based on that biological information. How do viruses comply with The Central Dogma?

It turns out that some viruses, like poxviruses and herpes, process their information just like cells. However, the vast majority of viruses that infect humans, such as coronaviruses, flu, Ebola, and HIV violate the Central Dogma because they have their genetic information in the form of RNA. (The red arrows below show how RNA viruses violate the Central Dogma.) Viruses also violate the Central Dogma because none of them have both DNA and RNA. Unlike cells, viruses either have DNA or RNA, but not both.[8] By the third definition, viruses once again fall short.

Viral Exceptions to the Central Dogma

So, by all three definitions, viruses fail!

Nonetheless, it is invaluable to look at viruses to see what they teach us about the rules of the Central Dogma of Biology and about life in general. Even though viruses may function in very unusual ways, in every case, viruses still need to replicate - they still need to copy their genetic information, they still need to produce messenger RNA as the instructions for making viral proteins, and they still need to be able to express that messenger RNA in the form of protein using the host cell's machinery. Because they resemble living organisms in these important ways, viruses are essentially "the exceptions that prove the rule of the Central Dogma of Biology."

QUESTION 2:

"To your knowledge, do all forms of life (prokaryotes, eukaryotes, etc.) have viruses?"

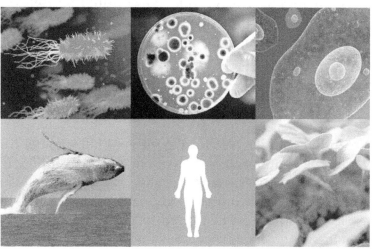

Forest Rohwer: (bio page 196)

As far as we know, everything has viruses. There are always viruses on viruses. It's that famous quote, "Every parasite has a smaller parasite." Lots of different mobile pieces of DNA carry on viruses and are parasitized by transposons, like virophages and so forth.

Nils Walter: (bio page 224)

It is very unlikely that there is any form of life that does not have a virus that preys upon it or interacts with it. This is because there is co-evolution between cellular life forms. The kingdoms are archaea, bacteria, and eukaryotic cells, the latter of which are cells that have a nucleus and cytoplasm, like the ones in our bodies. In archaea and bacteria, you have the two combined. You only have one compound and one membrane around the entire cell.

The viruses that we see today are relatively modern in origin and have evolved together with these cells. As soon as there was the first cell, probably there came to exist the first virus that would have infected that cell. Every so often, the virus will succeed in killing the cell, but if it does that, then its host will die, and the virus itself needs the machinery of the host cells to survive as a parasite. As a result, the virus will also vanish.

In the end, it's best for the virus to kill some of the target cells but not all of them. In that way, over billions of years, viruses, single cells, and multicellular organisms like the human body have co-evolved to co-exist. Due to the ancient origin of viruses, it's unlikely that there is any organism today that doesn't have a virus associated with it.

Richard Allen White III: (bio page 209)

The thought here is about ciliates. We've never found a virus in ciliates. I go back to when I was a graduate student at the University of British Columbia with Dennis Lynn, who was an absolute brilliant scientist and protistologist. He studied ciliates his whole life, this really bizarre group of eukaryotes. He was the man! I miss him terribly. He was brilliant! He gave some of the most amazing talks about an organism you've never heard of. As far as we know, they have a non-canonical genetic code and are not infected by viruses. Whether ciliate virus exist is unknown.

Richard Jacobs:

From what I understand, all forms of life do have viruses, although some haven't been discovered, for instance in some parasites and archaea. Even some giant viruses have virophages, which are viruses that infect other viruses,[9] and that's pretty amazing.

The lines appear to be blurred when it comes to viruses; plant viruses act very differently than animal viruses, which act differently than bacterial viruses, which act differently than the viruses of fungi, which are very likely to act differently than the viruses that affect parasites.

When considering all the different kinds of viruses that exist versus parasites, bacteria, animal cells, and plant cells, it becomes clear that there is an overlap of abilities between all of these forms of life (and yes, I do believe viruses are alive). Even the definition of a virus itself appears to be blurred a bit amongst these forms of life.

There are single-celled organisms that often don't exist in singular form, but in multicellular form, such as bacteria that exist as biofilm—as a collaboration of different cells. There are also holobionts, of which humans are a prime example; we are not made of just human cells, but of different tissues and organs and all the associated cells, not to mention trillions of bacteria, fungi, viruses, and the viruses that infect those bacteria, fungi, and viruses. A human is a huge compendium and collaborative effort of many different forms of life. Perhaps we hold every single form of life there is—viruses, bacteria, animal cells, virophages, fungi, proteus, and probably even archaea and (temporarily) plant cells.

If you think about the way viruses inhabit various organisms, the definition of an organism itself gets blurred; not only do we have all these creatures inside of us, but we have viruses that entered our cells long ago and became part of our DNA.

It's rumored that about eight percent of all our DNA is viral,[10] and a lot of this viral DNA is essential for us to live and be placental mammals. We also have viruses like herpes, which can go latent within us for entire lifetimes, or only come out during stress. Are these viruses separate from us or a part of us?

There are papers on various cancer tumors that appear to be a very different set of cells in our own cells, so they are yet another form of life that is inside us. There are bacteria that actually inhabit and become part of our cells.

Hundreds of millions of years ago, mitochondria appeared to merge with prokaryotic cells to make eukaryotic cells, or to provide an energy production facility. Mitochondria are an endogenized bacteria that merged with our cells,[11] so where is the line between bacteria and us?

In plants, the chloroplast is an endogenized blue-green algae that became part of the plant. Ultimately, it may be hard to answer the question of whether all forms of life have viruses, because the lines are blurred.

QUESTION 3:

"Why do some viruses multiply inside their host and lyse the host cells, while others integrate (retroviruses) into their host genome, and still others may go dormant or latent for a period of time and flare up only when the host experiences stress or deterioration?"

Eugene V. Koonin: (bio page 194)

This is a question for which there is no single answer. Virologists — especially those who study temperate bacteriophages — often talk about lysogenic choices or decisions. I think this is a very nice way to describe the behavior of viruses because they "make that decision" based on the state of their infected cells. If the cell feels good, viruses tend to be lysogenic and propagate within that cell. If the cell is under stress or likely to die, then the lysogen or the prophage will be induced and the virus will make the lysis choice, try to kill the cell, and escape as soon as possible.

Adolfo Garcia-Sastre: (bio page 214)

I think it has to do with diversity and evolution. Once something gets into a particular line of evolution, it is very difficult to get back—and this is true not only for viruses, but everything that has to do with evolution. Why do some animals need to eat other animals in order to survive, and some animals just need to eat plants in order to survive? Why do you even need to eat plants, when plants are living organisms that don't need to eat anything that is alive? Plants can survive just by having oxygen, sun and soil, because they can make energy out of oxygen and soil nutrients.

It has to do with biodiversity and how different forms of life have found different ways of behaving through evolution. For example, we as humans cannot just go in the sun in order to get energy. Why not? In theory, it should be possible if we would have the appropriate metabolism, but it's not possible because of the way we have evolved. It's impossible to go back to a time when an organism could make the split decision in evolution to get energy from the sun or by consuming other things.

St. Patrick Reid: (bio page 205)

I think it has to do with host adaptation and its introduction. As a biologist virologist, I studied viruses like Ebola. I think a lot of the pathogenicity that we see from these viruses is usually due to their new introduction. Relatively speaking, they were newly introduced to the host, and because they were newly introduced, they just replicated and killed.

One of the hallmarks of Ebola outbreaks is that the host gets killed so fast that it doesn't allow for the virus to replicate and spread; if you are a virus, this isn't beneficial to you because it doesn't allow for you to stick around. This is why we are called spillover hosts. I think the longer you are around a community or engaged in a process of seeing something, the more you are going to adapt to want to be there, which means that you won't want to kill the host too quickly.

For certain viruses, I think that it's the new introduction to hosts that allows for the virus to be more lytic. The difference between the viruses that will integrate and the viruses that have processes that allow them to stick around (e.g., herpes viruses, DNA viruses) has to do with the different nucleic acids (single-stranded RNA versus double-stranded DNA). With that said, we've learned from the West African outbreak that the Ebola virus can actually grow latent as well,[12] which brings a whole new paradigm to RNA viruses because we've always thought that the DNA and retro viruses are typically among the viruses that go into the nucleus and become latent.

We've always assumed that RNA viruses (like Ebola) that replicate entirely in the cytoplasm won't stick around; either the cells die, the host dies, or the virus leaves and keeps moving.

The West African outbreak showed us that there were a number of individuals who were infected with Ebola who seemed to clear the virus, and then months down the road, the virus came back.[13]

Michael Betts: (bio page 223)

There are a lot of different answers to that, because every virus is wanting to do something different. Ultimately, the goal of the virus is simply to propagate, and whether it does so at the cost of the host or not kind of depends on the virus, in some sense.

Some viruses kill you because they are replicating so quickly or they are pathogenic for other reasons, and others don't kill you or make you sick at all, yet they are still propagating. We actually have endogenous viruses in our genomes that we've lived with for millions of years; whether they were previously pathogenic, no one really knows.

Richard Jacobs:

As I mentioned in a previous answer, I believe that viruses are very much alive. When they enter a new host or a new environment, I believe there is a period of forced adaptation in order for the virus to survive, propagate, and either set up shop latently, or multiply itself, lyse the cells, and continue infection.

Once they've undergone this forced adaptation, depending on how long they are inside the host, there is going to be a forced re-adaptation over time, and this can change the nature of the virus. A virus that is first pathogenic may monitor the immune system and decide that it's time to go latent and hide out for a while because conditions do not support the ability for that virus to multiply itself; this is a forced re-adaptation.

Viruses can go through many different types of forced re-adaptation over time inside of a host, so there is no wonder that viruses can change behavior. With that said, some of them don't seem to be able to, as in the case of viruses that are purely pathogenic.

What we do know is that some retroviruses can endogenize into our DNA and go silent, or endogenize and remain active, like HIV. The herpes virus can be pathogenic and then go latent for days, months, weeks, years, or lifetimes. Some viruses have the ability to be latent, some have the ability to be pathogenic, and some have the ability to switch between pathogenicity and latency or even commensalism or mutualism.

Viruses must be able to monitor host conditions, both inside of cells and outside of cells, and they've been known to change their actions based on threats from the immune system and threats from other invading viruses or bacteria that try to enter the cell once they are inside of it. They monitor the availability of raw material, such as food, glucose, and other materials, as well as intracellular chemistry to ensure that the conditions are not hostile to them.

There are interferons and other chemicals and cell defenses inside cells that try to destroy viruses, and signaling on the outer membranes of cells which alerts the immune system to which cells are corrupted and need to be eaten. There is programming that causes apoptosis, and the virus has to monitor the conditions of its host cell to prevent apoptosis and to accomplish its other aims. If the cell has certain customary jobs, the virus may need to co-opt those for its own purposes. It must continually monitor and adapt to the changing conditions, both inside the cell and outside the cell.

There can be inbound extracellular vesicles that can give information to the virus about the environment outside of the cell. I read a paper on the topic of virosomes, which addresses the ability of some viruses to customize extracellular vesicles and have cells put them out to notify other cells of what's going on within that cell.[14]

Because of this, I believe that there is viral quorum sensing that happens amongst other infected cells. In other words, a virus is able to sense how many other cells in the environment are infected. In order for viruses to have all these behaviors, I believe that there must be incredibly sophisticated monitoring of both intracellular and extracellular conditions.

QUESTION 4:

"Why is there a latency period between first infection of a host and observable pathogenicity?"

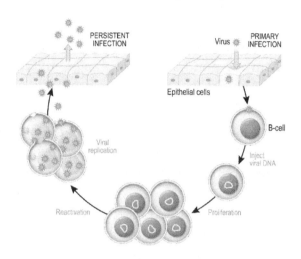

Denis Noble: (bio page 198)

Let's consider what we mean by a "latency period." We don't mean that the virus is not there because clearly it is, so it isn't that it takes time for a virus to get in. What is actually going on? If the virus just reproduces, then it's not going to do any damage. The question that must be asked is this: what is happening during the period that leads up to the pathology, and what is the pathology?

The pathology is not caused so much by just possessing the virus or viral replication; it is caused by the pathological response of the organism itself. For example, the real problem with COVID-19 is that it's provoking the immune system into some very

47

extraordinary behavior that leads the immune system to cause great problems for the host. It's the host's own immune system that is going into overdrive.[15]

The great majority of those who are infected by COVID-19 survive and do pretty well. They may have a short period of difficulty, a bit like a flu, but unless the system responds with a major inflammatory response (which is the major problem with the respiratory tract), it will be mild.

In summary, I think the latent period is more a question for the physiology and pathology of the reaction of the organism than an intrinsic factor of the virus itself.

Jeremy Barr: (bio page 217)

I find the latency period really fascinating because it's almost like a metamorphosis, similar to a caterpillar changing into a butterfly. You go from a structural virion persisting in the environment to a quasi-molecular state where the virus exists purely as genetic information, controlling and manipulating the cells. It's a huge and rapid metamorphosis of that lifestyle.

As to why and how they exist in that period, there is a huge range of reasons and mechanisms. I think it all comes back to optimizing the replicative success of the virus, not only in the bacterial host that it is in, but also in relation to viral ecology.

There are an incredible number of ways that the phage can drive and manipulate a cell, whether that's producing virions or giving the cell benefits to increase the virus's own chance of persisting.

Robert Siegel: (bio page 227)

In my teaching role, I make a point to disambiguate the term "latency", because we use it in three different ways in virology. First, we use it to explain the period of time between when a virus enters a cell and when it starts producing infectious viral particles.

Second, we use it to explain the period of time between when you are exposed to a virus and when you develop symptoms; that is often termed "latency," but it's more correctly referred to as the "incubation period", and I think that is what you are referring to here.

The third way that the word latency is used is in conjunction with certain viruses that can "go to sleep" inside cells. A great example are the herpes viruses like herpes simplex and chickenpox which can go into the state where they basically become quiescent and wait for months or even decades, for the right time to replicate. Most viruses cannot enter a latent, but certain viruses are really good at it.

Let's get back to the question of the incubation period. Why do viruses have an incubation period and why is it different for different viruses? The main reason viruses have an incubation period is that they immediately start replicating. First, there is a lag time between when they enter the cell and when they start producing viruses, and then there is a lag time between when there are only a few viruses and when there are lots and lots of viruses.

Generally speaking, viruses are doing a lot of damage from the very beginning, but it is on a microscopic scale. We don't experience any symptoms until the amount of damage or the effect of the virus is large enough to enter our consciousness. Many viruses can replicate for a while before we become aware that we are ill.

There is a second reason for the lag, which is that many of the symptoms that we experience — for instance, when we have the flu — are in fact due to responses by our immune system. Our immune system takes a while before it can respond to being infected.

When there are only a few viruses there, the immune system is not in high gear, but when there is lots of viral replication, the immune system is in high gear and we start feeling the effects of some of these immune mediators.

Why do different viruses have different latency periods? What we see over and over again is that viruses are selected for their ability to replicate and produce more virus particles. It turns out that viruses use many different ways of making a living and carrying out the replication process. Just like people have lots of different jobs, there are lots of different niches and strategies that viruses use to replicate and produce more virus particles.

Some viruses replicate very fast because they can outrun the immune system, but others replicate slowly so that they can be stealthy and remain below the radar of the immune system, which is also an effective strategy. I think it's important to look at the individual virus and try to understand what strategy that virus is utilizing in order to produce more virus particles.

Eugene V. Koonin: (bio page 194)

The trivial answer is that in order to cause symptoms or pathological changes, you need a lot of the virus, so it takes time. An organism is usually infected by a small number of virus particles, so it takes time for them to propagate exponentially. During some period of the infection, they propagate exponentially, more or less.

That's the general answer, but there are more complicated issues. For instance, disease emerges in and viruses infect only certain types of cells. In the case of poliomyelitis, they infect neurons and start their propagation with different types of cells, like intestinal cells.[16] It takes time to go through the organism and its different barriers, etc.

Nathalie Gontier: (bio page 219)

A team at the Weizmann Institute in Israel found that there is some form of quorum sensing. Viral particles can apparently communicate between one another about whether or not it is an opportune time to start the lytic phase. This is a form of biochemical communication, which allows them to communicate about whether or not to attack the bacterial host.

This raises a lot of questions, because it involves an entity that you can't really call an organism because it does not have that metabolism. At the same time, there are groups of viruses that, at the molecular level, are able communicate about not attacking the host or remaining dormant.

Richard Jacobs:

Classic Darwinism will say that there is a latency period because it just takes time for the virus to exponentially replicate and infect a number of cells sufficient to make the host feel sick. I think it's deeper than that.

Viruses appear to be capable of using cellular machinery for their own ends, and they can send out virosomes (i.e. customized extracellular vesicles) to inform other cells of what's going on. In addition, they are certainly monitoring inbound extracellular vesicles and intracellular chemistry and conditions.

In addition, they can apparently do viral quorum sensing to determine how many infected cells there are in the nearby area, and whether there are enough for a viral swarm capable of becoming pathogenic; if the conditions aren't right for pathogenicity, perhaps the virus remains latent or changes its strategy. I'm completely anthropomorphizing, but without doing so, I wouldn't be able to consider these possibilities.

Viruses can modify the surface proteins and immune-recognizable structures on cell surfaces in order to evade the immune system, buy time for them to replicate, and allow them to stay in a persistent latent state.[17] Viruses can pretty heavily modify a cell's activity, metabolism, surface features, and extracellular vesicles. In almost every way possible, viruses seem capable of taking over the cell and using it for their own ends, like a car that they modify, drive around in, sleep in, etc.

Another reason for the latency period is that it takes time and experimentation to adapt to a new environment. There is an initial forced adaptation when viruses first infect a cell, which may take time. Upon initial infection, the virus might not yet have the ability to make progeny, and there may be some learning involved. I do believe that viruses, in their own way, can learn and adapt, and that adaptation takes time. It takes time to set up shop and monitor host conditions.

I believe that there is early competition amongst quasispecies of an infecting swarm of viruses. In other words, it's less likely that it's one virus infecting one cell, and more likely that a whole bunch of viruses infect a cell at once. For instance, the "ABC" virus would be considered all the same virus, but doesn't have all the same nucleotide sequences, so they are a quasispecies. Each one of these quasispecies of the viral swarm likely has different abilities.

As a result, there may be an early competition amongst these quasispecies to successfully take over the cell's controls. There may be one of one kind, 50 of another kind, or 1,000 of a different kind, and the one that's 50 is superior to the others at taking over the cell. Once this process happens during the latency period, the virus can go about its other job of replicating more virions and lysing the cell.

Latency might also be explained in part by the virus being overwhelmed by early host defenses (e.g. the cellular production of interferon, immune system defenses). The early stage of infection may be like a battleground until adaptation or a somewhat steady state has been reached, or until the virus has taken over the cell's abilities and successfully hidden itself from the immune system.

All of these processes may take time — whether hours, days, weeks, months, or years. For all of these reasons, I think the explanation for a latency period goes beyond the time it takes for exponential replication to occur.

QUESTION 5:

"If I catch virus 'XYZ' from its source and label myself as number one (i.e. the first to be infected with this virus), and then I pass it on to someone else, who passes it on to someone else and so on, what would you expect to happen to number 100 in the passaging chain? In this context, the term 'passaging chain' is defined as the number of times that the virus has been passaged amongst organisms of a given species before it infects the organism we are talking about. Would you expect less virulence, more virulence, a move towards commensal or mutualistic behavior, or endogenization, and why?"

Paul Turner: (bio page 201)

This is where it gets really tricky. Let's consider viruses that have a pretty high mutation rate. In a train of transmission from one to 100, one would expect (although this might be a naïve expectation) the virus to get better and better at infecting the host. It will change over time such that the 100th person in its virus population is going to look substantially different than patient number one, because the virus will have evolved to be better at whatever it needs to be better at in that host.

If you consider a virus system that has a pretty high mutation rate, there might be a particular variant of that virus that is good for the actual transmission step. Within the host, a lot of growth occurs, the viral load expands, and mutations that might let the virus better interact with host cells may rely most on that key variant or key genotype at the transmission event, and it kind of resets itself.

It's able to do this because the mutation space is accessible. It has a high mutation rate, and even if that mutant that is great for transmission is not very prevalent during the peak of the infection in that person, it can reset back to that variant.

As far as I know, this seems to be true in HIV. There is a lot of variation in HIV within a human, but the tendency for some key genotype(s) to be the best ones at the initial transmission and infection of a new host stays fairly constant through time.[19]

I think that demonstrates that there is fluidity in viruses, and that they become very mutationally diverse. Most people think that viruses are changing all the time, which is true, but they can also go back to the starting point a lot easier than DNA-based mutational systems.

They have access to the old variant in a way, because they just mutate back to it, and there can be this toggling back and forth so that there is a constancy, if you will, over time. The different variants are important as a virus is transmitting from person one to person 100, and there might not be as much variability as one would expect.

Forest Rohwer: (bio page 196)

Across the board, the rule is that they attenuate and get less virulent over time. If the virus is really virulent, it either disappears or attenuates over time. That's just the rule. They get better and better at avoiding the immune system, and they usually become more temperate.

The ability to spread is much better by being less virulent and existing under the radar. In humans, when there is no indication that someone is sick, the chances of the virus moving to someone new are much better than if the person who is carrying the virus is really sick. It's just that we are good at doing that with our behavior, and that's going to be true of any animal. It doesn't have to be conscious; it could be anything that gives an indication that someone is carrying a virus.

Güenther Witzany: (bio page 210)

Mainstream science says that there are a lot of mutations in the genomes of viruses — especially RNA viruses.[18] From my perspective, this is the wrong approach, because mutations need error replications. Mutation is defined by replication that produces an error from the master copy, and I don't think this is the correct description.

This is from the last century, when mechanistic and parasitic viruses were viewed as poor molecules that would have to produce a one-to-one copy in order for there not to be a mutation; the understanding was that if the copy is slightly different than the master copy, then it's a mutation, which means it's an error replication.

From my perspective, this is rubbish, because groups of RNA stem loops constantly produce new sequences and bulges, which are single-stranded RNA loops that open the possibility of connecting with other single loops of other RNA networks. This is a very creative moment, because they can invade new organisms and produce new capabilities which they then give to the host.

In turn, this enables the host to adapt better to the circumstances of the changing environment. I would term this 'innovation competence' of RNA stem loops; it cares that the evolution of organisms on this planet may continue and adapt to changing environmental circumstances. This innovation competence is clearly a completely different narrative than error replication.

An anthropomorphic view is this: a poet who writes a second poem based on his first poem does not create an error in replication, but innovation and creation itself. The masters of genetic material — the viruses and viral-derived relatives — create innovation, changeability, adaptation, and capabilities which cannot be explained by errors. To describe this as error is to use the wrong description from last century.

Richard Jacobs:

There doesn't appear to be a term for this, but I'm going to call the passaging of a virus from one creature to another the passaging number. Some viruses appear to start out and stay pathogenic to their host, but these viruses will likely have low passaging numbers because the host will not live long enough for the virus to spread, let's say 100 times, and have a high passaging number.

I would say that most viruses become less virulent over time, because they are adopting host-based strategies that cause a host to house a virus for weeks, months, years, decades or even a lifetime. Viruses that can endogenize into their hosts seem to do so quickly, but over time lose their full abilities to escape the host genome. They lose parts of their genes and genetic makeup that allow them to do everything as an independently functioning virus, and they become a permanent part of that host.

Humans have what are called HERVs—Human Endogenous RetroViruses. Endogenization occurs not only in humans, but in many other animals and likely in many other creatures. Many HERVs have been with us for millions of years and thousands and thousands of generations, so if they were pathogenic, we wouldn't be here, because they would have killed off our ancestors a long time ago.

I would expect that if a host survives a passaging event, then by virtue of surviving, the virus has adapted its behavior to long-term infection, reducing its virulence just enough to use the host as a springboard for more consistent long-term spread; going latent for periods of time or becoming less virulent or even going silent.

Alternatively, a particular host has developed some immunity to the virus, which stymies its virulence and forces it to adapt by hiding from and evading the immune system or going latent for periods of time.

If a creature is infected and its passaging number is high, let's say 100, then I would expect a large amount of variation and adaptation to a more commensal or mutualistic state with the host. Again, this is because a pathogenic virus is unlikely to keep its host alive long enough to achieve a high passaging number—even a passaging number of 100 would be very unlikely.

Even in the case of HIV, with this somewhat cooperative co-dependent existence of the virus in order to maximize its replication and spread, over time an endogenous virus is acted upon and changed by its host in such a way as to become more mutualistic or beneficial. This is because the host's goal is to stay alive, maintain homeostasis, thrive, and procreate, and to do so, cells and hosts can often use or mold parts of viruses for their own benefit.

If an endogenized virus continues to be pathogenic, it would compromise the host's ability to procreate and to pass the endogenized virus on to its progeny. If a virus is truly pathogenic, then it is unlikely to allow its host to survive long enough and stay in good enough shape to procreate.

After 100 passaging events, for instance to inheritance, it's highly likely that an endogenized virus will become much more efficient, lean down its genetic sequence, and shed unnecessary parts that are metabolically expensive, since this would be the best path forward to its continued existence by virtue of a mutualistic relationship with its host.

If the host itself adapts the virus or adapts to the presence of the virus in such a way as to minimize the negative effects on the host (which the host would undoubtedly do in order to stay healthy), then the host would probably utilize the endogenized genetic material of the virus for its own benefit, such as by making useful proteins.

We see this in placental mammals; the placenta itself is composed of viral-like material that endogenized in placental mammals millions of years ago, and allowed for placental mammals to evolve.[20] In other words, without endogenized viruses, we wouldn't exist, because mothers wouldn't form a placenta.

QUESTION 6:

"Would you expect the genetic material of virus 'XYZ' in the previous question to be drastically changed by the time it reaches passaging number 100?"

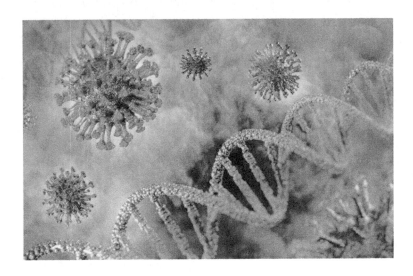

Gareth Brady: (bio page 221)

I suspect it doesn't change. It depends on the pressure, because there has to be a selection pressure to a great extent. There has to be some sort of selective pressure to maintain the mutations that happen, let's say, in a small RNA virus. On the topic of population sequencing of SARS-CoV-2, the number of people infected is probably in the order of hundreds of millions, if not billions.

The fact that it hasn't changed substantially in so many people indicates that there is no selective pressure for those mutations that are probably regularly happening to be maintained. That being said, viruses can change given enough selective pressure.

Viruses are most deadly when they've jumped from other species.[25] SARS-CoV-2 has likely come from bats, and is at its most dangerous because it is very poor at switching off the systems in humans (this is one reason it causes such profound inflammation).

You would expect over time that it will evolve gradually to be less dangerous. Most viruses just want to produce as much virus as possible while flying under the radar and not making the host too sick to move around and transmit the virus. And of course, dead hosts can't transmit a virus.

It's not the agenda of any virus to kill a host. It is usually the case that a virus that has recently jumped a species will find itself in an environment that it can't control or orchestrate. Over time, it will definitely change. When most people develop immunity to it, perhaps by directly getting a vaccination, any subsequent versions of it probably won't be dangerous anyway, because people will be cross-protected by having antibodies.

This is one reason most common colds in winter are viruses, and not that dissimilar from this one (SARS-CoV-2). However, they are different enough to the extent that common cold viruses don't kill people because they have adapted to humans, and over time, they become less dangerous as well. It's a process of adaptation that all viruses are evolving towards in order to maximize replication and spread.

Michael Betts: (bio page 223)

Basically, it totally depends on the type of virus, its genomic structure, the fidelity of its transcriptional machinery, rate of propagation, etc. Viruses like HIV or HCV, which have very high mutational ability would very likely show some degree of evolution after passaging through 100 people. However, it also depends on what you mean by 'drastic'; even with a high mutator like HIV, the evolution is much <1% over time. However, even that seemingly low degree of variation is actually very high, and can lead to marked differences in immune recognition.

Luis Villarreal: (bio page 203)

The capacity for persistent inapparent viruses to cause harm in other infections never goes away; it's an inherent selective feature that they maintain. For example, when it comes to mouse hepatitis virus in a population of mice that is persistently affected and showing no disease, the virus retains the capacity for causing disease in other populations. It will not lose this capacity with continued persistence and continued passage,[21] and I don't think that's going to happen with COVID-19 either.

These viruses are undyingly dynamic in terms of the composition of their RNA. Although there is one consensus type, which can be rather stable and is the most abundant version of the sequence, you always have lots of variants. As soon as you passage, you generate a quasispecies that has a lot of minorities and variants. Most of these variants don't have any phenotypes that we are aware of, but they are interacting with the population; sometimes they interfere and sometimes they complement, but they do all kinds of things within the context of that population.

There is evidence that the quasispecies is dynamic as it moves through tissue in one individual infection. For example, mouse hepatitis virus has the capacity to come in through the gut and move through nervous tissue.[22] These experiments were done with poliovirus in order to test the importance of this quasispecies dynamic in the emergence of disease.[23]

Since poliovirus has been studied a lot, it's possible to make a version of the polymerase that has high fidelity and a much lower error rate. Then, you can turn it back into a virus and use it to infect a mouse and study the resulting pathology.

With polio, the pathology that is particularly interesting is the neuropathology which spans the spinal column and gets into the central nervous system. It was observed that if there was a high-fidelity polymerase, the virus was unable to truly replicate far beyond the initial site of infection.[24] It did not seem to be adaptable to the neurological infection, and disease would follow.

Nils Walter: (bio page 224)

Ultimately, your question is about mutations. The virus needs very specific, very special properties in terms of how likely it is to infect, how fast it is to replicate, how easily it can recognize the next cell, etc. This poses a lot of restrictions on how much it can mutate.

For example, unlike HIV, coronaviruses have a proofreading replicase, which is an enzyme that corrects errors that it makes during replication. HIV has what is called reverse transcriptase, which has a tenfold or so higher error rate; it works because the HIV virus has a pretty short genome of under 10,000 RNA nucleotides, compared to 30,000 for coronavirus.

Since HIV has a shorter genome, it can make more errors without making so many errors that it loses the message. Overall, HIV mutates pretty rapidly, and spike protein gp120 allows it to evade the immune response and still infect cells.[26]

Coronaviruses, by comparison, make fewer mistakes intentionally because they have longer genomes. They can't make too many mistakes, or else they would wipe themselves out. While errors still occur, there are restrictions in terms of where they can occur.

For example, a famous G to T mutation occurred in Europe, and when the coronavirus came from Europe to New York City and the East Coast, it could be found there as a mutation.

As far as I understand, it's actually neutral in terms of the protein sequence; it's a mutation that affects the genome and may change replication speed, but doesn't fundamentally change the virus itself. This mutation allowed people to trace where the virus was spreading across the US and where it came from, so it was a great tool for tracing infections.[27]

There are relatively few mutations that occurred on coronavirus initially, because in order to be such an insidious and successful virus, there have to be many constraints on the spike proteins, with which it has to recognize the ACE2 receptor on a host cell, and all of the genes that are essential for its replication. These genes can't be changed too much without losing efficiency and the ability to be stealthy.

There are some mutations that will occur, and there is evidence that this has happened via the way the virus spread from China to Europe to the US. There is evidence that it might have become even a little bit more insidious after it mutated in Europe, but it's still limited in the scope of mutations that it can tolerate without losing its skills. However, with infections skyrocketing so have opportunities for the virus to mutate because of the sheer number of viral genomes being replicated. This shift is one reason why we need to get infections under control.

Richard Jacobs:

I would expect the genetic material to change substantially in the initial passaging events as adaptation kicks into high gear, and then slow down over time as the sweet spot of adaptation is reached. If a host experiences significant external environmental pressures, then at low passaging numbers I would expect the virus to revert to a selfish existence and turn lytic, like rats leaving a ship.

At high passaging numbers, I'd expect significant adaptation, which would make the virus's presence more permanent, and these stacked changes could possibly become irreversible as change upon change in the original infecting virus's genetic material occurs. This will make the viral sequence quite different over a high number of passaging events.

QUESTION 7:

"If I catch a virus from someone who nearly dies from it versus someone who barely experiences any symptoms, will I get very sick?"

Paul Offit: (bio page 211)

I think that you would likely be sicker, because people who are sicker are likely to shed greater quantities of virus. When kids get nascent disease, those who have moderate to severe disease shed more virus. In fact, they shed more virus compared to adults who get moderate to severe disease.

The good news is that children often don't get moderate to severe disease. I think there is definitely an inoculum effect, and I think that it does relate to the degree of symptomatology and the person who is shedding.

Shervin Takyar: (bio page 229)

There is a simple answer to this, which is that if the person is sicker because of their lowered immune status or because of higher replication of the virus, then they may have more viral load to give you.

Gareth Brady: (bio page 221)

There is a very strong correlation between virus load and exposure. In the COVID-19 situation, it was observed very early on that medical staff in particular were getting very serious infections. It seems they reasonably have evidence now to show that the virus load that you are exposed to at the beginning, and how effectively you inoculate all those viruses at once, determines how grave the disease actually becomes.

This makes sense, because if there is a lot of virus infecting you at the same time, it will reach critical mass faster, so there will be a much more dramatic response. That being said, the shedding periods during an infection differ between viruses. With SARS-CoV-2, people can be tremendously contagious at a very early stage before they even get a spike in fever.

Michael Betts: (bio page 223)

It depends on the nature of the virus, but I don't think that you can immediately predict. COVID-19 is the perfect example, in that we have no idea why some people have asymptomatic infection and others get sick and die. There are some associations with different genetic polymorphisms and disease severity for many different viruses, and there is some emerging information about SARS-CoV-2 as well.

In your scenario, the virus has a display of being completely apathogenic or pathogenic, so you can't tell. If it was always pathogenic, then you would probably have pathogenesis, but I think it's going to be difficult to predict. This doesn't mean it's forever impossible to do so. Certainly, there are genetic polymorphisms in the human population that will predispose you to be more susceptible,[28] particularly if you have polymorphisms and various immune response-related factors; that can be very problematic, and therefore a predictor of pathogenesis.

Paul Turner: (bio page 201)

That's part of my research, and it's called gene-by-environment interaction. If you have a virus and it dictates the disease state, is it a property of the virus and its genome, or is it an interaction between that virus and host A, and how sick host A is? If the same virus enters you as host B, will it dictate whether you become very sick or not?

Based on my intuition and my experience in research, I am a bigger believer in that the property of the virus to cause disease or not is a property of the virus, and it has more to do with the host's physiology or how sick they'll be due to other things that are going on in that host.

I guess that's one way for me to answer your question, but you can rip up and throw away that explanation if there is actual genetic change happening in the virus and host A before it reaches you. If there is a different kind of evolution in host B before it reaches you, then I guess I'm back to my original assertion that the virus has changed genetically and its interaction with host physiology will create a new set of rules.

Richard Jacobs:

It depends on the viral titer of exposure, meaning if someone has a tremendous amount of virus in them and you are exposed to them, it's more likely that they are going to pass along a significant amount of virus with their breath, sweat, blood, or other bodily secretions, whether it's transmitted sexually, by air, by touch, or by blood.

It also depends on the nature of the exposure. If someone coughs into the room and an hour later you come into the room, that would probably expose you to a lot less virus than if they coughed in your face, or if blood was taken from that person and literally injected into you, or if they bit or scratched you and there was a direct transfer.

I believe that the higher the titer of the virus and the more you are infected with initially or at a given period in time, the more likely it is that the virus will have a chance to overwhelm your immune defenses and make you very sick.

It also depends on your health status, immune system, epigenetic marks, genetic makeup, microbiome, and hormone and other biomarker levels.[29] In addition, it could depend on your constituent virome and phageome, because some of the bacteria, fungi, yeast, proteus, phages, and/or viruses that make up your microbiome and your body may have countermeasures to the virus that you are exposed to. This may make you more immune, or in the opposite sense, it may make it more likely that a virus would be able to make you very sick. It just depends on your makeup and the makeup of all the creatures that make you up.

If a person who was extremely sick infects you, then it's probably more likely that the makeup of the virus that made them very sick will dramatically sicken you as well, given that the virus was successful at significantly sickening them, although their makeup is different from yours.

If the person who infects you is barely sick, the same argument makes sense in the opposite way; you should experience a similarly low level of sickness because the virus, by definition, was not nearly as capable of making that person very sick. Of course, it depends on all the factors that make you "you," and the other person "them."

QUESTION 8:

"How much does my individual genetic makeup, epigenetic marks, immune history, etc. affect the degree to which I get sick if I get exposed to a virus?"

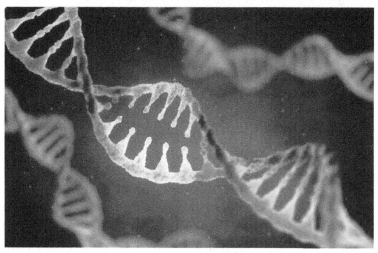

Güenther Witzany: (bio page 210)

The clear difference in behavior is lytic versus non-lytic. The lytic cycle kills the host, and the non-lytic cycle is persistent and does not kill the host. The difference may depend on the context within which the host organism is involved. If the host organism has a strong immune system which is able to regulate competing genetic parasites within it, then nothing will happen.

If the organism has problems with stress, damage, or a bad disease, then the persistent viral inhabitants may get out of control and become lytic again. It depends on the context in which the living organism is involved, not the mechanism.

We have some persistent viruses which will not cause disease because they have been integrated in the host organism for so long that the counterbalancing mechanism is very strong. The HIV and AIDS virus is dangerous for everyone who is affected by it, but some people are immune to it.[31] Investigations into this showed that in these people, the HIV virus has integrated in the host genome and is counter-regulated in a way that it will not harm the host.

However, if the host has sexual contact with another person who does not have the virus integrated in a persistent way, then that other person will become infected and may die.

In this way, it might become the case that over the next 100 to 1,000 years, many people will be immune to HIV. However, it will still be the case that if those people who are immune come into sexual contact with people who are not immune, the virus will be just as dangerous as it is today.

Paul Turner: (bio page 201)

I think that monitoring the host's condition is something that we underappreciate. For a virus that is in the cytoplasm or nested in the genome, I think there are plenty of ways for feedback to occur. For example, if the cell is stressed, feedback can make it to be interpreted by the virus as a trigger for them to destroy that cell and move on to another one. The probabilities of that happening, I believe, should increase as cells become more stressed.

One of our unpublished data sets talks about that in the bacterial world, where you can track a bacterial lineage as it is dividing over time; it's not immortal, but will eventually senesce or become incapable of making daughter cells.

We observed that when a lysogenic phage or a temperate phage is in that environment, it will be replicated alongside the host and reach the daughter cells, but the likelihood of it killing its host and leaving it through lysis increases not just through the arrow of time, but depending on how stressed or aged those cells are, which is itself a stressor.

We are still trying to figure out the mechanistic reasons, but at least in our own experiments, there is evidence of these feedback mechanisms, which I always equate to rats leaving a sinking ship. If the cell isn't working for you, then get out. I think it is underappreciated that viruses can somehow sense the cellular environment to do this, and I am excited to eventually publish that data set and see what people think.

Forest Rohwer: (bio page 196)

You have latent viruses everywhere and they are popping out relatively often. If you look at the blood of a healthy person, there is still a fair number of viruses in there,[30] and that's because there is always some proportion of them that is being activated and producing virions that can be shed.

When the system starts to become immunosuppressed, for example in the case of a metazoan, the viruses will start popping out everywhere and re-entering the lytic cycle. This is well-known in the case of HIV, and it happens in cancer patients all the time. Anything that will lower the health of the metazoan will lead to a whole bunch of viral production.

Richard Jacobs:

Your vaccine status may help you or hinder you. Vaccines are supposed to give you immunity to certain viruses, but they can cause a virus to find a different pathway to enter your cells that is actually more pathogenic.

Since the flu virus changes so often and so rapidly, you may have gotten a flu shot that protected you from last year's virus, but doesn't protect you at all from this year's virus. Vaccine status may help or hinder you, and it's not really known which one will happen.

Previous exposure to a similar or the same virus may also cause the virus to adapt and find a different, more pathogenic pathway, or it may block its action and thereby provide immunity, which is what everyone hopes.

By your genetic makeup, you may produce or not produce, over-express or under-express various proteins, enzymes, or other immune-protective molecules. You may have mutations, deletions, insertions, or other genetic alterations that predispose you to health or sickness.

Epigenetic marks are adaptations that occur over a person's lifetime exposure to their environment; they may upregulate or downregulate various body functions and genetic activity, and may predispose you to health or sickness.[32]

Your diet, sleeping habits, comorbidities, age, gender, race, previous medical history, the supplements you take, and the medications you use may all be significant factors. This is by no means an exhaustive list of factors that modulate how you will experience a virus, and there is no way to know how all these factors boil into whether or not you will get very sick from a given virus.

QUESTION 9:

"Why do viruses appear to have matched tropisms and subsequent host-to-host transmission methods? Examples: Rabies causes an infected host to bite other creatures, and spreads via saliva; influenza affects lung tissue and is spread by coughing and sneezing; HIV or HPV use sex to spread, and not coughing, sneezing or biting."

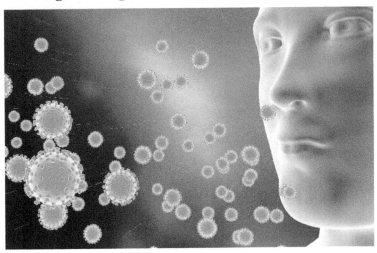

Shiraz Shah: (bio page 212)

There's a lock-and-key kind of thing going on, and I think this is because viruses have the ability to evolve very quickly. This means they can respond to changes within the host after just a few generations. Those changes could be the host trying to defend itself against that virus, and all kinds of other changes as well.

I think the tremendous ability for viruses to basically mutate means that they're also able to respond to their environment much faster than most other organisms. I think this is why you get these interlinked pairings where it almost looks too good to be true.

Why does that virus know to infect respiratory tissue? I think it evolved very quickly to do that. For example, adenoviruses cause respiratory infections, but they can also cause gastrointestinal infections.[33] Adenoviruses fall within several subfamilies, or you could say that there are a few species that infect human beings.

Some of the subspecies will infect human respiratory tissue and others will infect the epithelial tissue in the gut. I think this is just a matter of surface receptors having mutated within a few generations to be able to recognize either type of tissue.

Marilyn J. Roossinck: (bio page 200)

You have to remember that these are all happenstance relationships; initially, it's just accidental. SARS-CoV-2 happens to have a protein that attaches to the H2 receptor in human respiratory cells; this was happenstance, or perhaps derived from its previous host, and we don't really know what that was yet. Because of that, the virus can infect respiratory cells.

If a virus doesn't have a way to infect a respiratory cell, then it doesn't matter how many particles are floating around in the air because you are never going to get infected that way.

Evolutionary events don't happen overnight. These viruses have been around for a long time. SARS-CoV-2 has only been in the news for a short period of time, but it's been in its animal host for probably thousands of years.

79

At least in some hosts, rabies does change behavior; it's a virus that infects the nervous system and can change the host's behavior to biting, but that's not the only way it changes behavior.

In fact, when a cow has rabies, it starts choking,[34] which prompts the farmer to put his hands in the cow's throat to figure out why it's choking; in this way, the farmer gets a nip on the hand and contracts the rabies virus. That is a different kind of behavior that rabies can induce which also aids in its transmission.

Luis Villarreal: (bio page 203)

There is a tendency for what you're saying to be correct, but it is certainly violated by a specific example. Measles is a good example of a virus that violates a lot of the above things. It enters the upper respiratory tract and causes Koplik spots on the back of the throat.

These are actually diagnostic of early-stage measles, but they eventually end up everywhere, including the lungs, intestinal tract, and CNS[35] to cause seizures. Most of the pararetroviruses, like distemper, are very close relatives of measles, and they do pretty much the same thing. Viruses to tend to adopt particular 'life strategy' that involves specific patterns of host tissue infection.

Forest Rohwer: (bio page 196)

It's definitely matched nicely and it's because you always need that positive feedback for evolution. It has to be something that makes one route of replication that is better than another. For a virus moving in the blood of humans, it totally makes sense to spend part of its life cycle in the salivary glands of the mosquito, and that's independently coming through a whole bunch of different directions. Matching between tropism and how the virus makes a living is definitely going on and is strongly selected for.

Richard Allen White III: (bio page 209)

It's not really a matching, it's just that the virus found a strategy, and their entry is specialized. If the receptor is on every cell, then they can infect all of those cells. If a receptor is only on a handful of cells, in the case of HIV, then it can only infect those cells with the right receptor. It just so happens that it is a cell type that is just big enough that it can replicate itself in perpetuity.

The ACE-2 receptor that the COVID-19 virus binds to is found in many different tissue types.[36] If your receptor happens to be on many different cell types, or there is a protease that can cleave your spike protein that allows you to enter many different cell types, then you can take the host out. Again, this is not what I think a virus wants to do to kill its host: it wants to replicate in perpetuity. Viruses just keep replicating and sometimes they kill their host in the process.

Death of the host is just complete accident. A virus jumps into a new host where the machinery hasn't evolved   to stop it the host dies. Ebola is a classic example of this, as it replicates very slowly in bat cells due to resistance. In our cells, it replicates so fast that the cells explode. In any other scenario, that virus would just replicate without killing the host.

There are weird viruses like Zika, which has been in Africa since 1950 and we haven't seen phenotypes of microcephaly. However, in the South American population, it has caused an increased microcephaly. I think we're starting to work out the cause which is likely immunological.

Again, we don't see high numbers of microcephaly in African populations, but we see it in South American Latino populations,[37] so there must be something different about that genes or immunological background that allows for microcephaly to happen.

I don't think viruses pick the route of infection initially. I think that SARS-like viruses became respiratory because that protein (i.e., ACE-2) just happened to be in the respiratory tract, and through natural selection and evolution, allowed for entry, and that's they specialized in that route of infection.

Richard Jacobs:

This is certainly not the case for all viruses. No behavior in the viral world is universal, which is something I learned firsthand in many conversations with virologists and by reading many papers.

However, if you look at influenza, it tends to enter by respiratory droplets via coughing, sneezing, drooling, spitting, etc., and it also exists via those same means. Rabies appears to be contracted by bites and infected saliva—by a literal mechanical intrusion into the victim's flesh, and then it subsequently causes that same victim to have the urge to bite and drool in order to help the virus exit by that same mechanism. HIV and human papilloma virus (HPV) are transmitted by sex and acquired by sex, dissimilar to the flu and rabies.

Again, this doesn't occur for all viruses, but there appears to be at least several instances of matched tropism to a subsequent way that the virus transmits to a new creature. When a virus infects a given cell for the first time, there is likely to be a matching of viral ability that allows it to enter and fuse, and the matching will occur between the viral ability and the surface proteins or antigens that allow it to fuse or enter.

This could have started out as a chance occurrence, or perhaps due to extracellular vesicles of the past that were released by another cell type that has the particular cell as a target. Those extracellular vesicles have adapted enough to not only enter the target cell readily, but also take over the behavior and machinery of the target cell. In other words, the extracellular vesicle material, which can be RNA, runs the show and usurps the agency of the target cell in a way that allows it to completely dominate that target cell.

My speculation is that perhaps this is the origin of some viruses. Hundreds of thousands or millions of successive infection events of a given cell type will likely cause significant adaptation and tailoring of the cellular processes of the target cell, which enables more successful infection of future viral progeny of that cell type. It also enables part of the knowledge base required by a virus to be successful in infection, co-option of the cell's machinery, creation of progeny, successful packaging of it, and successful dissemination of progeny through lysing the cell and another victim picking up the virus.

This requires multiple adaptive steps, and because a virus must become an "expert" in cellular, mechanical, metabolic, and genetic works of a cell (i.e., how proteins are assembled, how DNA is copied, etc.), a virus's knowledge base will be heavily influenced by its target cell, which is its environment. The target cell will likely be used by that virus to complete the last step or all steps of its life cycle. The last step is dissemination of its progeny, not only within the host but outside of that host to a new host.

Flu viruses tend to infect respiratory cells and then adapt to disseminate through action of those respiratory cells and action of the whole organism by causing it to cough, spit, drool, snort, sneeze, etc., and thereby spread that virus.

A virus spends a lot of time with a target cell type and gets to know that target cell type extremely well. It will likely use tactics that not only allow it to enter the cell, co-opt the cell, and create progeny more effectively, but especially disseminate that progeny to the rest of that host and outside of that host to a new host.

QUESTION 10:

"Viruses are tools used by bacteria and other cellular organisms. Yet, viruses also use bacteria and other cells as tools for their own use. They do this as part of their entry, co-option of cellular machinery, creation of new genetic viral material, and through packaging into capsids and disseminating of viral progeny throughout a host and to a new host. What governs who uses who and to what extent?"

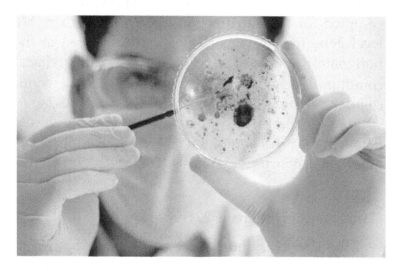

James Shapiro: (bio page 202)

I think if we look at virus-cell interaction as part of a complicated system, and assume it's the system as a whole that is operating, then some of these interactions make more sense because they allow the system as a whole to develop in certain ways.

Viruses do carry DNA that proves very useful to bacteria. It wouldn't be in the interest of a bacterium, for example, to exclude all viruses, because then it wouldn't be able to take advantage of the DNA coming from the viruses. I don't know if these are conscious decisions, but this is the way that the system evolves most efficiently.

Eugene V. Koonin: (bio page 194)

It's a really a major shuffle of the evolution of life, that as you pointed out very correctly, viral cells are co-opted for cellular function. For instance, there are local gene transfer agents which are involved in the horizontal transfer of genes in bacterial and archaeal communities and vice versa.[38] Viral cells recruit cellular defense systems for their own purposes, so to speak. Oftentimes, it is done to cope with the host's defenses, and it is part of what is often called the arms race between viruses and hosts.

Gareth Brady: (bio page 221)

I think it's very much a battle between the two to survive. The fastest evolving genes in all animal genetics are innate immune genes,[39] and that's primarily the reason. You can really see it happening when certain populations are slightly more resistant to an infection because of an abrupt mutation which gave them a preferential survival advantage over another population. There have been a couple of examples of this over the years, and that doesn't really happen with anything aside from innate immune genes.

Michael Betts: (bio page 223)

Being an immunologist, one of the most fascinating types of viruses — even more than HIV — are the herpes viruses. This is because they've evolved to carry copies or analogs of our own genes that they have modified in a way that allows them to evade or modulate immune responses.[40]

For example, herpes viruses have analogs of cytokines, which are proteins that our immune system uses to essentially marshal responses from other cells. An infected cell will start to express viral proteins that are copies of the analog cytokine (which is just slightly wrong), and this steers the immune response in the wrong direction.

It's really quite fascinating that this has evolved, and there are multiple examples of it. Poxviruses also do this type of thing. The bigger the virus is in terms of genome size, the more it has co-opted and included aspects of the immune response as a way to subvert the immune response and perpetuate itself.

Richard Jacobs:

If you take the position that viruses are alive and that all living organisms have a drive towards homeostasis and reproduction, and have some level of ability to adapt, then "who" becomes a battle of adaptation by the hopeful tool-user and the tool victim itself.

In some cases, viruses adapt to the point that they can regularly and reliably use infected cells to multiply themselves. In other cases, cellular defenses can be effective to the point that the infecting virus itself becomes co-opted by the target cell and becomes a tool of that cell. An example of this is in Vibrio cholerae; a portion of bacteriophage genetic material becomes inserted into the bacterium's genetic material and unlocks an ability to make the bacteria more pathogenic to a human host.[41]

Whether the bacteria have the agency to utilize the virus's genetic material and abilities or the phage itself has the agency to change the target bacteria for its own ends in response to us, for instance, being yet another host, is unknown. There are many examples of viruses, bacteria, fungi, animal cells, and other cell types that make tools of other cell types and become tools of other cell types. No one knows the parameters of this negotiation or of this war, if either metaphor is more accurate.

QUESTION 11:

"What viral cell entry method fascinates you and why?"

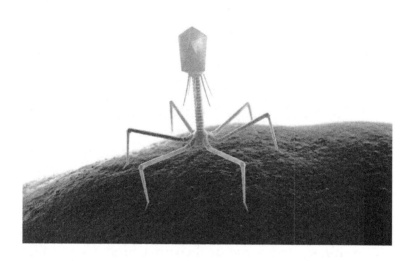

Paul Turner: (bio page 201)

I'm fascinated by the very many ways that viruses can enter cells. The thing that impresses me about some bacteriophages is that they act more like viruses of animals in the sense that they sometimes have envelopes that fuse with the exterior of the bacterial cell, which allows them to physically interact with the surface and be taken up. That's pretty wacky in the phage world.

A lot of the other phages that are especially well-known and well-characterized will sort of sit outside the cell and inject their genetic material into the cell. I'm impressed by RNA viruses called Cystoviruses, which are phages that get entirely engulfed within the cell.

Shiraz Shah: (bio page 212)

Maybe it's an error on my part, but I tend to think of these biological systems—especially viral infections—as pretty dumb. Much of the behavior that looks almost intelligent is basically a numbers game. If there are millions upon millions of viral particles, then some of them are going to end up doing something that ends up being clever because there are just so many.

With bacteriophages that replicate within the bacterium, you get 100 bacteriophages, and only a few of those need to mount successful infections in order for that bacteriophage to multiply. Even if most of those bacteriophages are wasted, it doesn't really matter. That's what I would think intuitively, but I still think you have a point. Biology often surprises us with these things, and some of these machineries are a lot cleverer than we thought.

Michael Betts: (bio page 223)

Every single one of them is fascinating in and of itself. How on earth could a particular set of circumstances be built upon to do what it does? HIV, the one I would certainly know the most about, is fascinating in the level of complexity that has evolved to not only allow entry of the genomic material in the cell, but to protect that virus from the consistent pressure of the host's immune response or antibodies.

Conformational changes have to take place, and only certain conformations are seen during the entry process. Otherwise, the immune response rarely sees these things and it requires multiple proteins to be involved. Every single one of them is fascinating in and of itself, to be honest.

Adolfo Garcia-Sastre: (bio page 214)

Most people, including scientists, think that viruses have very smart mechanisms to enter cells, replicate, etc. I might be one of the few scientists who believe the opposite; I think viruses are extremely dumb. They are just what they are because that is what they have been evolved to be. They look like extremely efficient machines, but they are very error-prone.

Most viruses — even the one causing COVID-19 — make a lot of progeny that are not able to survive, so it's just through sheer numbers that viruses achieve success. If they are able to make millions and millions of copies, then some of them will be successful, but many of them have a few mutations that will make them unsuccessful. They waste a lot of energy, but they are still able to do it.

The mechanisms how viruses replicate and take over a host may look very sophisticated, but I think they are due to evolution, and they aren't really so sophisticated when you think about it. If you think about how to make the most efficient organism that is able to propagate from cell to cell, you would probably choose different mechanisms than the ones viruses have. I think that viruses are dumb, but they are very efficient. You don't need to be very smart to be very efficient.

Richard Jacobs:

The entry mechanism of T4 bacteriophage that infect Escherichia coli is amazing to me. First of all, the phage itself is very complicated; it's shaped like a moon rover or an alien, with a head, a collar, and filamentous legs that drift over the surface of a target bacterium's exterior membrane.

These phages find perches, and then another set of legs comes out, centers, and locks in the positioning of the phage. The phage is then drawn close enough to the membrane of the cell for the bottom of its collar to touch the cell membrane, and then the conformation of the collar changes and opens.

Once the phage fuses, it screws its way down through the membrane, and the opening of the collar is then wide enough to allow for the injection of the phage's double-stranded DNA into the target cell's cytoplasm.

To be able to accomplish this sophisticated series of events properly is a remarkable feat of viral engineering. It is because of hyper-complicated and sophisticated phenomena like this that biological organisms appear to act in a deliberate, adaptive way versus a random way. They are not passively waiting for natural selection and random chance to allow them to live and do what they do to proliferate; they are using deliberate successive adaptation.

QUESTION 12:

"Why do some phages have a head, tail, and tail fibers, while viruses that affect eukaryotic cells tend to have a centralized capsid (hexagonal, spherical, etc.) with radially distributed spike proteins? What about rod-shaped viruses or filoviruses? What benefit do you think the various structure types have?"

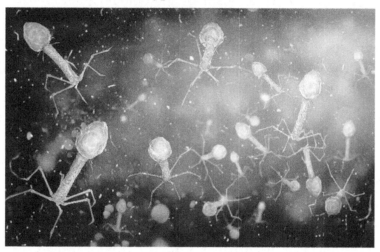

Joseph R. Masci: (bio page 216)

There are tremendous differences in the sizes of viral genomes, from very tiny to quite substantial. Some agree that the size of the genome dictates the shape of the virus. I think there are certain shapes that accommodate long nucleotide strands better than others, but this may be an indication that viruses didn't all emerge from one progenitor, but came from a variety of different directions and environments. They certainly occur in a tremendous variety of sizes, particularly of their RNA or DNA.

It's an interesting observation that they come in many shapes and sizes. The typical icosahedral shape is the standard shape that is often given for a virus, perhaps because flu viruses have that shape. The shape of viruses like Ebola and SARS-CoV-2, which don't have that seemingly rigid structural shape, could be dictated by the size of the genome, or something related to their environment, metabolism, or ability to invade.

We just have to keep in mind that there are about 300 viruses that infect humans that we know about, and that is probably a tiny fraction of the total number of viruses out there. To draw conclusions from what we know about it may be a bit treacherous. I also think that there are a lot of viruses that we currently don't have any way of culturing or looking at under an electron microscope; we only have evidence of their transmission.

St. Patrick Reid: (bio page 205)

Ebola is a filovirus, and filo is Latin for thread. The Ebola virus looks like a thread and sometimes comes in what looks like a knot. When it comes to shape and what drives the release of viruses, there are specific viral proteins that will recruit the host membrane and lead to an enhanced membrane curvature as it is being released. The shape is dictated by viral proteins that will help it to bud and release. Different viral proteins will allow for different curvatures to exist. But with that said, it's still not a fully understood process.

Entry is directed and driven by the proteins on the outside of the virus. SARS-CoV-2 has a spike protein, whereas Ebola has a glycoprotein. They are all spike proteins, but the name for the spike protein on influenza is hemagglutinin (HA) protein. The proteins will bind to a host's receptors on the surface of the cell, which will trigger entry into the membrane of a cell.

It turns out that Ebola doesn't necessarily have a receptor on the outside of the cell. The current knowledge is that Ebola gets engulfed by macropinocytosis,[42] which basically means that the cell sucks it in. Once the cell sucks it in non-specifically, it enters an endosome. Once it's in the endosome, it sees its receptor.

The glycoprotein of Ebola virus is the outer protein that allows entry, and is about 120 kilodaltons. This protein has to be cleaved by proteins called cathepsins down 19 kilodaltons allowing it to be able to bind its receptor. This means that it goes from being huge to really tiny. Once it becomes really tiny, it recognizes the receptor, which is in the endosome and is called NPC1. It's an entirely convoluted, complicated process.

Shiraz Shah: (bio page 212)

It is such a good question, and one that has been puzzling virologists for as long as I can remember. We're starting to see some answers because we're starting to get to know more and more viruses. About 90 percent of all viruses that infect bacteria look like moon landers, as you mention; by far, most of the viruses that infect eukaryotes are round and don't look very complicated.

Why do viruses that infect bacteria look the way they do? Two bacterial viruses that appear physically identical can have genomes that look completely different, so there is speculation about whether convergent evolution is going on, which means that, perhaps, the only way to infect a bacterial cell is to look like a moon lander. There have been many hypotheses as to why this is the case.

Almost all bacteria have a cell wall composed of something called peptidoglycan, which is a polymer that is very thick and hard to penetrate. At the moment, the speculation is that the reason most bacterial viruses look like this is because they need to be able to penetrate this wall. Archaea and eukaryotes don't have a peptidoglycan cell wall. Cell walls in eukaryotes can be composed of cholesterol and other molecules which are much softer than peptidoglycan, but they can also be composed of harder materials, like cellulose in plants.

There's a lot more diversity in the composure of cell walls within eukaryotes and archaea, so it seems that there isn't one dominating structure in terms of the variants that infect eukaryotes and archaea. We think that almost all of the viruses that infect bacteria look the way they do because almost all bacteria have the same type of cell wall.

Denis Noble: (bio page 198)

As a physiologist, I'm fascinated by the fact that we classify the cell types in our body into at least 200 different types.[43] Some cells secrete calcium to form a hard bone, others send out filaments the length of the organism itself and transmit signals to each end, and still other cells perform metabolic tasks specific to the liver. The interesting thing is that all cells come from exactly the same genome, which we all inherit from our parents.

By a process of differentiation, there can be an absolutely phenomenal range of outcomes. For this reason, I think it would be surprising if viruses did not find a way of developing variation. It seems to me that organisms do it so much already — especially large multicellular organisms like you, me, and many of the animals and plants that we know. They have the ability to interpret the genome in very many different ways.

Another interesting topic is that viruses must also use their hosts to change. You see, if you can't replicate outside a host, then how does SARS become a coronavirus? It must be that replication inside hosts leads to a change in the virus; this is the only way in which a virus can change.

Richard Jacobs:

Some would say that a long series of random mutations is responsible for the tremendous variation in virions, but since my underlying assumption is that viruses are alive despite their virion stage appearing to be a static, non-motile, and unchanging one, I believe that countless previous successful infection events inform viruses of how to adapt better to their host's cell targets.

In some cases, this means a round capsid. In other cases, there will be a sophisticated moon lander type of structure as seen in the T4 E. coli phage. Some viruses have a capsid and a second membrane-like envelope, some don't. Some have radial spike proteins, some don't.

In working effectively with a target cell shape and morphology, host morphology, the local environment of target cell types, and the type and number of exterior cell membranes, as well as the cell's defenses, surface proteins, and other structures, viruses have co-evolved through successive deliberate adaptation to continue to successfully infect their targets, some over millions of years and trillions of successful interactions.

There appears to be no single body plan of a virus, nor any one single entry method into cells. There are dozens or more of each of these phenomena, and they appear to be dependent on a myriad of conditions as I have described above.

QUESTION 13:

"If you de-nucleated a cell or sucked out its contents and left its membrane intact, would a virus that normally infects it still complete its fusing and entry, or would it stop partway through, sensing something was "wrong"?"

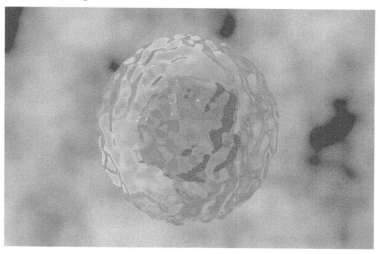

Richard Allen White III: (bio page 209)

It would definitely enter. It would eject its components (depending on the virus), but that would be the end of it. I think it would just do its thing, and just sit there.

Eugene V. Koonin: (bio page 194)

I am unaware of any evidence of such sensing. I think that if the membrane receptors are intact, then the virus would go inside the cell, only to find out that it was not worth it.

William B. Miller, Jr.: (bio page 206)

The answer lies in whether or not viruses have a Senome. The Senome is a new concept that František Baluška, at the University of Bonn, and I have talked about with respect to cells. Cells have a Senome, which we consider to be the total assembly of the sensory apparatus of the cell.[45] It is everything it can deploy to receive and assess information. I don't know whether or not a virus will enter a damaged cell or not.

However, if it had a Senome, it probably wouldn't enter, because it would know that there is an advantaged habit for it to go where there are more readily available intracellular resources. No one knows this answer, but a sharp focal point can be placed on that analysis. The focal point of the question is really whether the virus has a sensory apparatus and knows its environment well enough to make contingent decisions.

Joseph R. Masci: (bio page 216)

I think that there are viruses that may be able to sense or detect the difference biochemically. If that could be used as a therapeutic strategy, it would certainly be interesting to consider. It also goes the other way in the sense that the formation of the viral genetic material and the formation of the viral capsid are not always linked with each other; there can be empty viral capsids floating around. I wouldn't be surprised if a bacterial cell could sense that a viral capsid is empty, or a virus could somehow detect that a bacterial cell is empty, but I don't know if there is data to support it one way or the other.

Jeremy Barr: (bio page 217)

My first answer is that it would absolutely adsorb and infect that "decoy" cell, because bacteria do produce small vesicles that don't contain any genetic information as a decoy to sequester phages.[44] The virus DNA would enter, and then realize there are no ribosomes and nothing it can utilize.

With that said, phages can mechanically sense their bacterial hosts. The T4 bacteriophages are my favorite viruses; they have tail fibers which mechanically wrap up and sit around the head (that's their highest energy state). Occasionally, they will snap down, reach out, and sense the environment. In doing this, they can actually sense whether or not a bacterial cell is actively replicating, dead, or a persisting remnant of a cell fragment. I'm not exactly sure of the mechanisms for this, but I believe they sense the electrostatic potential on the cell.

Richard Jacobs:

The answer to this question hinges upon whether you believe that viruses are alive and have some level of ability to adapt to their environments and host cell targets. Although virions appear to be non-motile and dormant, similar to a chicken's egg or the seed of a plant, the moment they encounter a target cell's outer membrane, certain viruses appear to awaken and exhibit behaviors that are unexpected, assuming a virion is truly passive and non-motile.

For instance, some viruses appear to make initial contact with the target cell's membrane and then re-orient themselves to be perpendicular to that membrane to begin the process of membrane fusion. Others have been noted to "crawl along" the surface of a target cell to find a secondary receptor that enables entry.[46]

I believe that certain viruses — based on their initial contact and entry mechanism — when encountering a denucleated or defective cell will perform a membrane-level sensing of the condition of the target cell, or will fail to encounter a needed receptor or an entry pathway. In either case, the entry of the virion will cease at that point. What happens from there? No one knows.

If the problem occurs late enough in the fusion and entry pathway, it may be irreversible and continue regardless, allowing the virion to enter. Or, the virion may be stuck to the cell membrane and stuck in an intermediate state until the cell dies or is predated by another organism taking the virion down with it. If the problem occurs too early (which appears to be what many virologists have considered as a method for blocking the entry of a virus), it may lead to a deliberate adaptation of the virus to find another, perhaps more pathogenic and successful method of entry and infection.

This may conserve the ability of virions to move on to other target cells, perhaps making infection of unblocked cells more likely, because more virions will end up surrounding, fusing, and entering them. It may be better to know all the steps involved in viral entry for a given virion and a given target cell, and then choose a later step that will compromise the virion, yet save the cell from infection, co-option, and destruction via lysis.

In asking this question, I wanted to put out the idea that it may be possible to create a series of decoy cell targets that have the proper outer membrane characteristics but lack the appropriate machinery inside the cells to allow a virus to create progeny after entering. The virus, in this case, would enter the cell, shed its capsid, and exist as a naked RNA or double-stranded DNA inside the cell where there is "no one home." The virus would be trapped and unable to multiply.

QUESTION 14:

"How similar or different are extracellular vesicles versus bacterial plasmids versus viruses?"

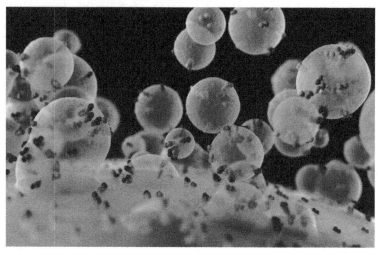

Lily Wang: (bio page 218)

The plasmid from bacteria is a very amazing kind of genomics information that can replicate, but it's not a virus. You might experiment with the plasmids you release that are manipulated to enter the cell. In nature, I am not too sure how the plasmid transmits through one bacterium to another bacterium. It should be a different mechanism than the mechanism of viruses, and the plasmid seems to be shared information between the population of bacteria.

The virus might have originated from some kind of plasmid, and information could have been transmitted between cellular life. Later, the plasmid-like materials could have gained the ability to produce the structural proteins and assemble themselves into a virus-like form, which would make it easier for them to transmit between the cells. I think that the original viruses were just simple plasmids.

Luis Villarreal: (bio page 203)

Those are all pretty related, and many bacterial plasmids are partially derived directly from defective viruses or parasitic viruses. Viruses indeed also have viruses of themselves. The situation becomes quite complicated as you look at it more in depth. Hyperparasites (i.e. parasites of parasites) and viruses make defectives, and defectives are derived from the viruses themselves to interfere with the viruses if they colonize the host[47] So this becomes a network which plasmids are part of.

Defectives dramatically change the whole host-virus's trajectory. The overall circumstance becomes exceedingly complicated once you start putting it into a system of life. Viruses are just part of a matrix of parasitic entities that are interacting in a network in ways that are much more complicated than we've historically considered.

If you try to separate what a plasmid is from what a virus is, I think you've missed the importance of communication and parasitization, which is what they both are. They are frequently derived from one another, and interact with and indirectly affect one another. They are a part of the same parasitic milieu that makes up life.

Eugene V. Koonin: (bio page 194)

They have features in common. In the case of plasmids, there are direct evolutionary relationships with viruses,[48] yet they are not viruses. However, it depends on the definition. The best way to define viruses, I believe, is to say that they are selfish genetic elements that encode structural proteins and encapsidate the genome. From that point of view, vesicles or plasmids are not viruses, even though they may be related to viruses and resemble viruses in many ways.

Richard Jacobs:

Extracellular vesicles (which I will call EVs for short), bacterial plasmids, virions and other intercellular exchanges have many elements in common. All contain payloads that include genetic material that, once inside a target cell, may cause changes in gene expression, metabolism, transcription of RNA, usurpation of cellular machinery and cellular agency, and more.

EVs, plasmids, virions, etc. all enclose and protect their cargo inside of a membrane or multiple membranes, or inside capsids or other protective structures. All of them have very specific cell targets and specify entry mechanisms that match the entering particles' morphology and structural/chemical abilities with the target cell membrane's receptors and other factors. All of these may enable reporting mechanisms of the target cell as to the success of the entry, and may be used for pure communication, co-option of the target cell, or deliberate change in the target cell.

To add to this list, we can consider humans and the dynamics of conception. Sperm cells are similar to motile virions, extracellular vesicles, bacterial plasmids, and other biological constructs in that they have an outer protective protein coat and/or membrane. They contain genetic payloads, and are themselves the target of inbound EVs or other biological entities, such as epididymosomes and prostasomes.

A select sperm can enter into an egg cell after communication with the target egg cell via EVs, or perhaps once the "right sperm" finds the appropriate receptor on the egg cell surface, it fuses, enters and successfully "infects" the target egg cell, co-opting its cellular machinery to change and direct its fate from that point forward and form a zygote. What does this tell you…that EVs, plasmids, viruses, sperm cells, and other cell types, and cellular communication and co-option tools have many similar characteristics? I leave that answer to the reader.

"Do you believe in the "one virus, one target cell" model, or are multiple viruses needed to co-infect?"

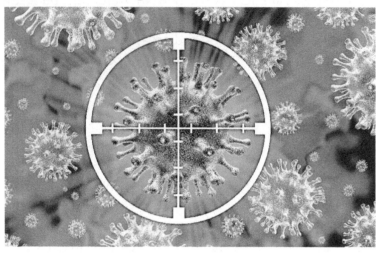

Matthew Frieman: (bio page 230)

One virion can easily infect a cell and cause disease and replicate. If there are multiple virions in the cell, then there would be a greater amount of genome to start with, so there would probably be more viral replication in a shorter period of time. Certainly, one virus can enter a cell, cause disease, replicate, and make a million particles quite easily. There is no requirement for multiple virions to be in the same cell.

Marilyn J. Roossinck: (bio page 200)

I know of examples in both plants and fungi where more than one virus is needed in order to infect. These viruses are sometimes called helper viruses; one virus provides one part of the function and the other virus provides another part. A lot of plant viruses and fungal viruses have divided genomes, and tend to package each genome segment in a separate virus particle.[50] Those are like individual components of the same virus, but if there are four or five different components, then they all have to get into the same cell in order to establish an infection.

Paul Turner: (bio page 201)

I would adhere more to the idea that if a virus or virion is able to enter a cell on its own, it will. I'm pretty skeptical about coordination across viruses being something that evolved for that purpose. Viruses often have a lot of variation, and if they are interacting with a cellular population — whether bacteria or the cells of the liver — then the variation can help them, depending on the way that they are interacting with those cells.

This creates complexity for the cells in dealing with the problem, and that might make them more vulnerable to one or more of those viruses being successful. I don't know if that's evolved for that purpose, or if it just helps viruses to more easily gain infections and be successful. It's a bit of a gray area, but I need more convincing that it's a coordinated effort.

I think it would be highly cool if there was the equivalent of the division of labor in viruses. A very popular model for the evolution of cellular life relies on the idea that a division of labor occurred, and that rather than one cell dictating its own fate, different cells have different roles and interact with one another through signaling, eventually evolving to multicellularity, which contributes to the success of cellular life.[49] The question is whether you really need that in order to be successful for life, and I don't know that you do, because viruses are the most plentiful things on the planet.

Lily Wang: (bio page 218)

I'm not sure if that exists in physiological conditions, but in lab conditions, we use very high titers of virus and infect a cell. Usually, it is a saturated concentration of virus. We mark one virus with green protein and one with red protein, and can see both of the viruses enter the same cell (because we can see both colors).

When one virus enters, it can downregulate certain surface proteins when infecting the cell. I think those proteins could be beneficial or detrimental in terms of the ability of the other virus to enter. There is a mechanism for the entry of the virus, and the pressure of a certain number of proteins will definitely remold the environment inside the cell. It might be beneficial for another virus and it might not be; it's complicated.

Jeremy Barr: (bio page 217)

Most of classical biology says that each virus is responsible for its own infection and its own entry into a cell. However, the most exciting thing about virology is that viruses break every single biological rule that we make.

There are examples of viruses that are completely dependent on other viruses to replicate. I gave an example of what we call a cryptic phage, which lives inside a bacterial host cell, but can't replicate on its own and has to rely on other viral infections to reproduce.

Another example is P2-P4 phage, where one of these phages relies on the other to actually cause the infection, package, and transmit[51] the viral particle. Both of these viruses must be used together to propagate.

Richard Jacobs:

Every scientist I've spoken with tells me that there are 1023 to 1031 viruses on earth, and that viral titer (the amount of virus that a person is exposed to) can vary from 107 to 1013 (106 is one million).

Since a person can be infected with tens of millions to quadrillions of viral particles, and since my guesstimate of the successful number of viral infections over all time is likely 1015 or more, it's extremely unlikely that a single virus reaches a target cell membrane and fuses to it, while countless nearby virions do not. I think it's a near certainty that successful infection of a cell requires dozens, hundreds, or thousands of virions to be successful.

Given cellular defenses to viruses and other attackers, viral entry is also likely assisted due to the slight or significant variations in phenotype of a given virus infecting a cell. When someone is infected by a virus, even if that virus has a name, there will be differences, mutations, or variations in the genetic sequence of the infecting virus, as viruses have a high mutation/variation rate. Not only will a wild type viral swarm have members of differing genetic sequence, but those members will have different capsids, envelopes, spike protein conformations, cell receptor affinity, and other differences, making it more likely that at least several variations of the infecting virus will gain entry into a cell and require it to defend itself against a range of attackers of different phenotype and abilities.

It is unknown whether virions, when in the process of fusing to a cell membrane, can signal one another via cross-membrane or cross-cytoplasm signaling (using the cellular machinery itself to signal other inbound virions as to the cell's condition and how to enter) to facilitate entry and infection. Perhaps inbound virions can even alter the biogenesis and packaging of extracellular vesicles (EVs) that are given off by a cell, which may be intercepted by external virions that interact with these EVs.

In addition, the initial infecting virions of a target cell may have enough time to change the expression of various outer cell membrane receptors, ligands, proteins, etc., which may facilitate or hinder entry of other similar virions of the same type of virus.

"Can viruses co-opt cells to send out altered evs, or do quorum sensing to determine whether there are other cells infected with the same virus? Once a virus co-opts the cellular machinery, can the virus use that machinery to communicate as a proxy for the cell in cell-to-cell signaling?"

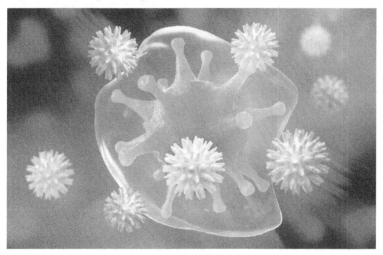

Robert Siegel: (bio page 227)

I don't know of any example where that might be true. My inclination would be to say that they don't do that, but again, the answer to almost everything about viruses is that if a virus can acquire an ability that facilitates its replication, then you would expect to see it.

What is more pertinent and similar to what you are talking about is the response of the immune system. The immune system definitely has the ability to have cells recognize that they are infected and to send out signals to other cells to try to protect themselves. Human cells and the human body are smart, but the virus has the advantage of being able to replicate and change very quickly. This is basically warfare, in which the body is being very deliberate about trying to get rid of the virus, and the virus is selecting for those variants that happen to be able to outwit the body's defenses.

Shiraz Shah: (bio page 212)

Rotem Sorek is a researcher from Israel who discovered a system wherein bacteriophages can basically coordinate infections between each other. A cell that's infected by a particular bacteriophage will create a signaling molecule which will get exported outside of the cell, and that signaling molecule will be picked up by other cells, which are also infected by bacteriophages.[52]

In that way, the bacteriophages can coordinate infection in order to make the infection more efficient. This is because the virus doesn't want to kill all the host cells at once; it wants to phase it out in order to ensure the long-term survival of both the host and the bacteriophage. It's amazing that this can happen.

Michael Betts: (bio page 223)

If you are going to invoke communication, then you are kind of asking whether the virus is alive. I don't believe that viruses are alive.

Bacteria and parasites can do things like quorum sensing, and both are live organisms. Viruses can affect their environment by changing the cell they are in. For example, some viruses will co-opt a host cell's protein machinery, causing the host cell to make viral proteins instead of its own proteins.

This might cause that cell to be sensed in its local environment in a different way because the cell surface proteins may change. It may become sort of de-attached from the adjacent membrane, but I don't know. I don't think that a viral infected cell would be talking to another viral infected cell saying, "Hey, let's do something together."

There are syncytia, which are cells that merge with other cells that merge with other cells, resulting in a big multi-nucleated conglomerate of cells, and viruses can cause that to happen.[53] In that sense, you can sort of imagine a scenario like you are talking about, where one infected cell meets up and merges with another infected cell, but they are not truly communicating; they are just kind of aggregating.

Shervin Takyar: (bio page 229)

Some very interesting information recently come out, particularly about how phages have a social life and talk to each other. It seems like during most functions carried out by viruses—including what is known as quorum sensing, which is a term that came from bacteria that use it—viruses do message each other, and these messages are translated through the cells' translation machinery (since we are talking about phages, this would be the machinery of bacteria). They use the machinery or the factory per se inside bacteria or inside a cell to translate their messages and talk to each other. This is very good, recent research that Dr. Sorek has published. It's fascinating.

St. Patrick Reid: (bio page 205)

We do know that cells secrete extracellular vesicles (e.g. exosomes) as a form of communication on a normal basis, so it stands to reason that when a virus infects a cell, it co-opts that system. As a matter of fact, some viruses will leave through exosomes and some viruses will manipulate what's being released from the exosomes. To me, this implies that the virus is actually manipulating the host's signal; whether it's manipulating the host's signal in such a way as to detect the presence of pathogens, I wouldn't speculate that because that requires a little too much coordination.

I think that once a virus like Ebola gets into a cell, it's just replicating. I don't think Ebola virus is getting into a cell and trying to identify whether there's another virus. I think that could only be the case if it could cooperatively allow for replication to enhance with time. If both of the viruses have infected the host over time and there is a beneficial relationship between them because the host allows them to both be there, then the signal that they will secrete will allow that to occur.

However, for that cooperativity to evolve, they would have had to infect the same host for eons. With that said, phages associated with the commensal bacteria in the human microbiome are probably doing that.

Let's say the virion in cell A makes a conscious effort to communicate with cell B. I would posit that the virion in cell A doesn't care about what's happening in cell B; that virion is secreting what it's secreting to help itself, and by helping itself, it releases things that happen to help the virus in cell B, but that is not its purpose.

Viruses aren't trying to quorum sense; they are just trying to replicate. If by chance, what the virion in cell A releases allows for the virus in cell B to replicate, then it would be a cooperative relationship that would allow for it to flourish. However, the virus is not doing it to help another virus; it is doing it to help itself.

Richard Jacobs:

An answer to this question wasn't known at the time of the interviews I conducted for this book, but I was subsequently given a scientific paper that describes viral quorum sensing.[54] This paper (A Host-Produced Quorum-Sensing Autoinducer Controls a Phage Lysis-Lysogeny Decision by Bonnie Basser, et al.) was sent to me by Charles Bernand, a PhD student at Sorbonne University in France. Thank you, Charles!

Another interview, conducted after the interviews for this book were finished, was with Jason Shepherd, who described to me a human gene called ARC. This gene creates and packages RNA into a capsid that is extremely similar to known viral capsids. This virion-like particle is then packaged into extracellular cargo and expressed by the cell.[55]

In addition, I have learned from conversations with various virologists that just as cancer cells release customized EVs containing unique cargo, viruses appear to have the capability, in taking over a cell's machinery, to significantly alter the cargo contained in their EVs as well.

Since one of the main purposes of EVs appears to be a communication channel between cells, as well as a method for cells to influence the gene expression and other biochemical processes of target cells, and since viruses appear to be capable of harnessing many cellular processes, it's likely that an infected cell (called a 'virocell' by Patrick Forterre, a French researcher, writer and biologist) can not only send out customized EVs, but perform quorum sensing by creating the right molecules to do so, emitted as cargo inside EVs or as naked molecules, or expressed on the infected cell's exterior membrane.

QUESTION 17:

"Ants, bees, wasps and termites are types of superorganisms. They are swarms of the same species of creature, with several different phenotypes, yet all acting for the same good of the colony. Animals, plants, and other organisms are also superorganisms in that they are assemblages of cells, microbes, viruses, etc. Bacteria often form biofilms, which are complex structures of one or more strains of bacteria, and can be reckoned as superorganisms in themselves.

Can viruses be considered to be a swarm or a superorganism, in that a wild-type infecting virus likely contains dozens, hundreds, or thousands of genetic variants (known in some circles as quasispecies) and infects in huge numbers, perhaps acting as a swarm organism?

Do viruses have a sense of 'self' vs. 'other,' or recognition of quasispecies as 'self,' and do they coordinate infection, latency, or other behaviors? If so, what are the implications of this line of thinking?"

St. Patrick Reid: (bio page 205)

I think there is something to that. There is the idea that during infection, some viruses will make different versions of certain genes, or make half a portion of their genome. Effectively, they make quasispecies during the course of infection, and the one that replicates the best is the one you see.

During the course of an infection, there might be different virions of the same species replicating. For example, Ebola might be replicating poorly, and so one protein might come out halfway, one protein might come out all of the way, and one protein might be mutated, but it's the same virus—just different versions of the same thing. The one that's the most fit is the one that will keep replicating.

To your point, this can happen with influenza in certain species. For instance, a flu virus that specifically infects birds can infect a pig, and the same is true for a flu virus that specifically infects humans. As a result, a pig could have both viruses and act as a mixing vessel.[58] This wouldn't result in a whole new virus, because it would still be a flu virus (albeit a different-looking flu virus).

William B. Miller, Jr.: (bio page 206)

Quasispecies should be considered an essential element of our evolutionary model. The same construct can be usefully applied to cells, though that term is not actually used in cellular life. I'm saying this because the quasispecies concept has been used as a theory for the origin of life. The idea behind this is that a series of primitive replicons was established before protocells or with protocells, and that this kind of mutant cloud, based within the quasispecies concept, is how we got to present-day RNA viruses and maybe present-day life itself.[56, 57]

I'll go one step further. Let's go to the macro side and consider ants. When you see ants exploring their environment to look for food and circling around in patterns that seem both non-random and random, it pays to remember that their actions mirror cells.

In effect, though invisible to us, cellular life is doing the same thing at its scale, we are always producing variants; that's the point of sexual reproduction, we are producing both stochastic and non-random variants (within limits) that can explore the environment and bring that information back to the organism. This is the living process that has been operating among the cellular domains for billions of years. So, it is proper to consider viral quasispecies as swarm form of environmental (intracellular) exploration.

Shervin Takyar: (bio page 229)

This is a very interesting question you bring up, because it has some philosophical implications about evolution. Let's talk about quasispecies. Quasispecies are different sequences of the same virus in a host. I worked with quasispecies when I was working with hepatitis C. We know how they are generated; there are polymerases that viruses and all other organisms use to replicate their genome. These polymerases can be permissive, meaning they are not high fidelity, and they allow some mistakes in sequence. That's how quasispecies are generated. The polymerase of hepatitis C is permissive, so there are quasispecies — many sequences with some mistakes or variations that come out in every host.

Why is this important? Quasispecies may work the same way that mutation works in evolution. There are types that may be more resistant to or more capable of survival. In that sense, quasispecies can be a tool for survival. Are the quasispecies in particular cooperating?

It comes down to how much you submit to the whole theory of the selfish gene, which is the philosophy that there is a particular sequence or a particular life form that replicates more of itself, and all the rest is a manifestation of that. You have probably heard about Dr. Wilson's theory regarding ants and bees, which is that different ants and bees are extensions of the genome of the queen; each has an expression of their own gene, but they are extensions of the queen.[59] In that sense, they are all working together to propagate the genome of that queen bee or queen ant.

Marilyn J. Roossinck: (bio page 200)

I don't think viruses have a sense of self because that's something that requires a higher order of consciousness, and I don't think viruses are conscious. The term quasispecies comes from physics and not biology; it has nothing to do with the biological concept of species.

It is a word that is used frequently for RNA viruses in particular. Some DNA viruses have similar behavior, in that when a virus infects a host and begins to replicate, it makes a lot of mistakes in its replication process, and those mistakes aren't usually corrected (some viruses can correct them and some can't).

This generates a very diverse population within a single host or host cell, which is a quasispecies. They do not act as individuals, but as an entity; the quasispecies is an entity. Some may be defective and some may have different functions in different parts of the genome because of the rapid evolutionary process by which they are making so many mistakes in the copying of the genome.

If you know anything about genetics, then you know that we generally have two (sometimes more) alleles of each gene. You can think of a quasispecies as an entity that has thousands of alleles. It has to be infecting, and it has to be a single replicating population. It's in one host and one cell, not spread out among a bunch of hosts; it's a single population.

Richard Jacobs:

I hypothesize that infecting viral swarms not only recognize 'self' versus 'other' in their action once they have infected a cell, but that the infecting swarm can identify the quasispecies of itself as 'self' and 'other,' with distinct viruses as the 'other.'

Amongst close genetic variants of virus, there will likely be a diverse set of abilities, some of which will make cell infection more likely, some that will allow for alternate receptor binding or cell entry methods to infect, some that will better counteract the host's immune system and cellular defenses, and on and on. In addition, the variability of an infecting viral swarm due to the varied capability of the quasispecies will lead to more successful infection.

I hypothesize that viruses alter their makeup after each successful infection of a cell, and that the resulting variants act as a collective "memory" of how previous infections were carried out successfully. This fits neatly with a portion of neo-Darwinism, in that "survival of the fittest" doesn't have to mean a single type of virion to which all others are copies; fittest takes on a new definition to mean the fittest swarm of variants, all achieving higher fitness to successfully infect due to their diversity of structure, action, and ability.

In addition to viral variants acting as memory, they act as a viral immune system, because some of the variants will be more resistant to cellular and host defenses. This is why I hypothesize that an infection by a viral isolate of only one genetic sequence will not be as successful as an infection by a wild-type virus.

The same scenario likely applies to vaccines; the efficacy of vaccines can likely be increased by creating vaccines that have many genetic variants, creating a more successful defense against a diversified, wild-type virus infection.

Viral swarms that successfully infect cells also likely coordinate lysogeny and latency behaviors.

"If a cell is infected, would the infecting virus alter the cell to prevent infection and entry by other viruses, like a dog guarding a bone?"

James Shapiro: (bio page 202)

I think there are examples of viruses that can downregulate the expression of receptors for other viruses, or produce proteins which might cover the receptors for other viruses. There are also viruses which have CRISPR systems, so when they infect a cell, they can protect the cell from infection by other viruses.[61]

Paul Offit: (bio page 211)

The virus doesn't do that; it's the host's immune system that does that. In response to an infection, your body will make interferon, which is a particle that interferes with the capacity of other viruses to infect at the same time.[60] This is the host's immune system's general non-specific response that has a general antiviral effect, and this is what will make it more difficult for a second virus to infect.

Richard Allen White III: (bio page 209)

I don't know of any examples of that. With influenza, definitely not; you can get infected by multiple virions at the same time within the same cell, this is how swine origin influenza outbreak of 2009 happened. You can also be infected with HIV and GBV-C at any time.

Forest Rohwer: (bio page 196)

Absolutely; this is the main thing that is happening. The cell is carrying provirus. In the majority of cases, I would assume that the virus is actually protecting the cell against other viruses and even helping it to hide. For example, if you are a virus inside a single cell in seawater, then you need to protect your cell against other viruses and against things that would eat you, like proteus and so forth.

It's really common, because viruses have to protect their cells or they are dead. It is probably the main selection pressure on cells, because it is so common. Every cell that you have has been under multiple viral attacks, so if you are a virus inside a cell, you know you are going to be at war to protect that cell.

Richard Jacobs:

There appear to be many examples of this in the scientific literature, where it is referred to as 'superinfection interference,' amongst other names.[4, 62, 63]

If viruses are alive, then they will have the same homeostatic, reproductive, coordinate, cooperative, and survival drives that other creatures do. It should be no surprise that, once a virus infects a given cell, it must defend the virocell's existence against cellular defenses, host immune system attack, other viruses that would attempt to steal its prize, and so on.

QUESTION 19:

"Once a target cell is infected with a virus, do you believe that the virus's action will alter the local microbiome that surrounds and interacts with that cell?"

Jeremy Barr: (bio page 217)

It will do this in many different ways. The first key point is whether you are talking about a eukaryotic virus that can infect and replicate within the mammalian cells, or another virus, such as a phage. I'll keep coming back to the phage perspective, but I can speak more broadly to the eukaryotic virus.

To give an example, a viral infection of the cells in your gut can change the metabolism and processes that are happening within those cells, which can directly influence the receptors and the molecules that those cells will display on their surface, and potentially even change their secretions.[64]

One of the big research areas that our labs are looking at is the mucus layer in the gut. This mucus layer is critically important for your gut microbiome as it provides a food source and a structural resource for these microbes to grow on.

There's been growing evidence that viral interactions — not only pathogenic ones but symbiotic or off-target ones as well — can actually change the mucus structure in the gut, which can change the growth of specific gut bacterial and viral species. This can have knock-on effects on the function of your gut via metabolism, inflammation, or any number of other ways.[65, 66]

James Shapiro: (bio page 202)

I would expect so, because microbiomes involve nutritional, signalling and other interactions between different cells and different types of cells. Virus infection will alter or disrupt those interactions and consequently perturb the mixture of organisms that is present around the infected cell.

Richard Allen White III: (bio page 209)

I think that if you have a massive viral infection or any kind of infection, or eat bad food, or are not healthy, it can cause dysbiosis of the gut. There is no doubt that if you are a runner or an athlete and all of the sudden you come down with the flu for two or three weeks, your microbiome will really change within those weeks. Eventually, you will get back up to speed.

In people, I think there might be viral infections that cause dysbiosis of the microbiome. Pushing their gut microbiome off the train tracks. What do I mean by that? I want to think of the human microbiome and human health as one train moving forward, but viruses, bacterial infections, poor diet, stress, and poor genetics might push the train off the tracks, leaving people with dysbiosis, which may cause early death. I think that more work, in general, needs to be done in order to look at all of these factors leading to viral infection and microbiome dysbiosis. I think it's a field that is ripe for amazing discovery.

Richard Jacobs:

I interviewed Florencia McAllister, who is studying pancreatic cancer tumors. She said that the pancreas has its own localized microbiome, and that the cancer tumors on the pancreas itself have their own localized, distinctly different microbiomes.[68] This leads me to believe that humans (and other creatures) not only have gut, oral, skin, vaginal, eye, and other microbiomes, but likely have microbiomes associated with every cell type and every niche in the body. Every single one.

The association of a local microbiome with somatic cells is governed by the trading of resources with the local cells, cell metabolites and nutrient needs, avoidance or cooperation with the local immune system, cellular membrane receptors and other identifying substances, etc.

Since cellular infection by viruses appears to change their profile of extracellular vesicle production and cargo, expression of their surface membrane receptors and other structures, and cellular metabolism (and who knows how many other changes), this would directly lead to a change in that cell's interaction with other cells and with the localized microbiome.

Perhaps resources that were provided by cellular metabolism lessen or cease under the direction of an infecting virus, or perhaps there is upregulation of certain membrane structures that were used to interact with that cell by its localized microbiome. The list of possibilities continues, but I strongly believe that a cell's localized microbiome will shift quickly after infection by a virus.

"Could the phageome be responsible for the immunity of bacteria (e.g. phages providing plasmids that are useful to bacteria in evading or resisting antibiotics, conferring virulence as in cholera, and forming adaptive responses to the environment"?

Robert Siegel: (bio page 227)

In the case of bacterial viruses, some of them harm the bacteria and some of them help the bacteria. If something benefits both the virus and the cell, then it will be selected for.

Cholera is infected with a phage called CTX, which stands for cholera toxin. That phage encodes a gene for a protein that causes massive diarrhea.[69] From the bacterial standpoint, that's advantageous to the bacteria; massive diarrhea might hurt the host, but it makes it more likely that the bacteria will spread to a new host. In this case, the virus benefits the bacteria, but in the vast majority of cases, viruses hurt bacteria.

Before 1920, a proposal was made that perhaps viruses could be used to kill bacteria and stop bacterial infections with the "my enemy's enemy is my friend" strategy. That was used before the development of antibiotics. With the development of antibiotic resistance, people are looking back at the idea of using bacteriophages as a kind of therapy for bacterial infections. This strategy is advancing fairly rapidly for people who have cystic fibrosis and for people who have bacterial infections that cannot be treated with currently available antibiotics.

Jeremy Barr: (bio page 217)

There are some incredible molecular and ecological mechanisms that phages use. One example is the superinfection exclusion mechanism. This is when a bacterial host is already infected by a phage (typically by a lysogenic infection), where the phage is laying dormant in its bacterial host and riding it out in the hopes of better days before it ends in the lytic cycle.

The superinfection exclusion mechanism will actually block similar phages from infecting that same bacterium.[73] From the phage's perspective, it doesn't want another virus coming in and stealing its host for its own use, so it will produce a range of proteins and enzymes that will either block the ejection or penetration of new phage genomes into the cell, or actively degrade those that are within.

It can also delay lysis events. There is a system whereby every subsequent phage infection event will be sensed by the cell and its infecting phage, causing a delay in lysis.[73] Through this mechanism lysis can be delayed for hours or days.

From an ecological perspective, the phage is infecting and filling its host cell with new virions, but it won't want to lyse that host if there are a huge number of viruses already out in the environment, because it would just be competing against them. So by continually sensing how many infections are coming into the host cell, it can choose to delay lysis until that cell is no longer being infected to optimize its reproductive chances.

Lily Wang: (bio page 218)

A viral infection of bacteria might lead the bacteria to store the information of the specific virus and then render it with new features against that kind of virus. It may also gather the information from the virus and introduce other features, like antibiotic resistance.

If you add antibiotics to a bacterial population and most of them die, only those with resistance would live, and I think those are the ones that have plasmids with antibiotic resistance. The bacteria that are left also have their plasmids propagated as more of the plasmid, and we can transfer this information to the neighboring bacteria. This would result in a whole population immune to the new antibiotics.

Richard Allen White III: (bio page 209)

There are definitely cases of phage having protective elements. Phage have a "double-edged sword," because they both giveth and taketh away.

There are examples of phage that infect bacteria called *Pseudomonas aeruginosa* on a wound, and in turn, make the bacteria more virulent, which prevents wound healing.[70] This increase in virulence is common, and can also be found in the phage of *Shigella* and E. coli thus taketh away of the sword.

The giveth of the sword is when a phage can protect a person from something like non- alcoholic fatty liver disease.

Shiraz Shah: (bio page 212)

Viruses that infect bacteria place an enormous pressure on bacteria, because every other bacterium gets killed by a viral infection every other day. Viruses are a lot more lethal to bacteria than they are to us, so bacteria are always on the lookout to create new defense systems to protect themselves against these viruses, which is accomplished through evolution.

The interesting thing here is that viruses have a higher turnover rate than bacteria, and bacteria have a higher turnover rate than humans. We live for maybe 80 years, and one human generation is about 30 years long. It takes 30 years before human beings typically have kids, but some of the fastest growing bacteria have a new generation of bacteria every 20 minutes. With viruses, it's about 100 times that number, because each virus will infect one bacterium, and during that life cycle of 20 minutes, it will create about 100 copies of itself. Viruses multiply 100 times faster than the bacterial hosts that they infect.

Not only do viruses replicate a lot faster than cellular organisms,[71] but they're a lot simpler than cellular organisms; since they're made up of fewer moving parts and components, fewer components are dependent on one another, which means that the viruses can experiment with the contents of their genome. As a result of that, they can evolve a lot faster than bacteria.

Why am I telling you this? The reason is because there seems to be some evidence that in this arms race between viruses that infect bacteria and bacteria themselves, there's a lot of evolution going on, and a lot of that evolution is happening on the viral side because the viruses replicate much faster.

Many of these viruses will integrate into the bacterial chromosome rather than kill the bacterium. When they become part of the bacterium, many of the protein-coding genes that are on the viral genome become part of the bacterial genome. This means that the bacterium is carrying around genes that encode viral proteins, and some of these viral proteins might actually help the bacterium survive. It's basically a survival strategy for the virus.

Sometimes it makes better sense to help your hosts rather than kill your hosts. Since there's so much evolution going on in viruses because they replicate so much faster than bacteria, there are always new kinds of proteins that are evolving from scratch. One of the leading hypotheses within the field is that most proteins evolve from scratch within viruses, and then they are inherited by cellular organisms to carry out regular tasks,[72] like powering life.

Richard Jacobs:

I believe so, considering the example of Vibrio becoming more pathogenic after insertion of viral material into their genetic material.

In the case of antibiotic resistance, perhaps this is partially how a bacterial immune system arises, given that the microbiome OF a bacterium (which may be a new concept) is largely composed of phages; some that kill, and others that bolster and assist adaptation of all sorts.

Bacterial viruses (aka phages) are libraries of genetic material and abilities, and bacteria seem perfectly capable of being infected and controlled by them, as well as taking charge and cutting out bits of useful "phage DNA" that they insert into their own genome, which gives them adaptive abilities to their current environmental, predatory, and host-interactive challenges.

QUESTION 21:

"Holobionts, organs, tissues, and possibly even individual cells have localized microbiomes to themselves. Do you think that the cell itself is an assemblage with its own microbiome, in which the RNA, DNA, and other materials within it are constituents of the cell but all have their own degree of agency? Contemplate cellular and/or viral agency."

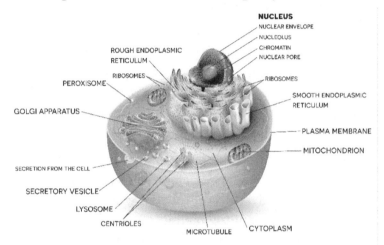

Nathalie Gontier: (bio page 219)

It all depends on how we define agency. James Shapiro from Chicago University says that you can find agency in mobile genetic elements in viruses, genes, and plasmids. He says that as soon as you have those elements, you can say that there is agency because there is some kind of functional behavior.

In a way, they know what to do and what not to do, and there is some kind of self-preservation. There is no way, at this moment in time, to define that kind

138

of agency; agency has been defined by a kind of 'self,' and that 'self' has been defined by a kind of consciousness, self-indemnification, self-defense, etc.

Genes, mobile animals, bacteria, and viruses don't have a brain or neurological system that enables them to have any kind of self-awareness. If we want to talk about cognition at that level, then we have to redefine it and take away any kind of consciousness or neurological basis of cognition.

There are a lot of philosophers who say that when you take away the neurological part of cognition, there is no cognition at all and you have to call it something else. Is there a difference between thinking neurologically and behaving molecularly and genetically?

Robert Siegel: (bio page 227)

In thinking about viruses, we are repeatedly impressed that something so simple can be so smart. Basically, the key to this riddle is molecular or evolutionary intelligence. It's like you have this little computer program endowed with the ability to replicate, but it doesn't do it perfectly every time.

If some of the programs become better at replicating, they will take over and pretty soon you won't see the original program at all; you'll see the modified program. From there, they can modify it further, and as they make each of these modifications, they get better and better at reproducing.

It looks like they have a strategy, but what they are really doing is just selecting for the variants that are most effective at carrying out the mandate of replication. The mandate is that if things can replicate well, they'll come to predominate and we'll see them. If they can't replicate well, they'll disappear from the population. There is no will or intention; it's just about selecting those variants that are best at this little game. You can think about viruses as essentially little biological computer programs.

Paul Turner: (bio page 201)

I've long been a fan or at least a proponent of the idea that different biological levels exist and that natural selection could occur in all of them. I think scientists and a lot of people in the lay public appreciate that genetic elements or other things that parasitize genomes have their own autonomy, if you will, or at least they have their own agency, as you stated it.

Why would we think that the selection acting on a transposon or prophage or virus is the same as what's happening on the host? There's a different outcome for selection depending on the strategy that you have to work with as a biological entity.

If you can see selection happening at the cellular level, there could be a different kind of selection happening at the sub-cellular level. That's my point. This could be coordinated in the right direction if both of them are working together and there is a net benefit or a synergy to whatever they are doing in their environment. If suddenly they are in another environment where that synergy breaks, then they could have selection acting on them differently and this agency could start to take over.

Forest Rohwer: (bio page 196)

I know that some people do think that, but I think of it a little differently. Any living thing — whether you want to call it an organism or not — has to find an environment where it can replicate. This is true whether you are a tree, human, cell, or virus. It just turns out that viruses tend to go where they can get all of the things that it takes to make a new virus; they can just steal these things from the environment very directly.

A tree does this in a pretty complex way, by converting basic elements and molecules into something that is living. If you look at it that way, it's really just a continuum of finding the right environment in order to continue through time. It's easy to put a human cell into a tissue culture dish and get it to replicate, but in that case, we are really just feeding it everything it needs. This can be done with viruses too; you can give them the internal workings of the cell, and in some cases, get them to repackage and make a new virus.[74]

Richard Jacobs:

It used to be thought that many places in the body were "sterile," including the bladder and the urine within it, the brain that lies within the protective blood-brain barrier, and in and around most of our organs and tissues. Within the last 20 years, scientists have discovered not only our gut microbiome, but dozens of localized microbiomes in our mouths (one under the tongue, one in the back of the throat, one at the gumline, one below the gumline, etc.), a skin microbiome, an eye microbiome, a vaginal microbiome, and others.

I hypothesize that every organ and every tissue in our bodies attracts, interacts with, and trades with its own localized microbiome, which is composed of bacteria, phages, viruses, protists, archaea, fungi, yeasts, parasites, and other organisms.

A lightning bolt struck me when I interviewed Florencia McAllister about pancreatic tumor microbiomes being different from surrounding, healthy pancreatic tissue. If tumors on an organ have a different microbiome than the surrounding healthy tissue, then this likely means that there may be no such thing as a single-celled organism.

Even single-celled organisms or single somatic cells may attract a small, somewhat diverse microbiome that stays in the proximity of that organism, knowing that there are resource trading opportunities, better communal/combined evaluation of the harsh surrounding environment, greater protection against predators, and a benefit of somewhat heightened immunity.

Let's take this thought further. If one believes that viruses are alive, and believes that viruses are only "alive" once they enter into a host cell... what has entered the cell that is "alive"? If a naked strand of genetic material can cause such substantial change inside a cell, and in some cases, completely co-opt that cell, then what of the various native RNA types (mRNA, tRNA, etc.), organelles (mitochondria especially), and other structures and entities (cell nucleus, our DNA itself) inside our human cells?

Where does identity of "self" stop? Could a cell be a holobiont unto itself, when considering its small, localized microbiome, the RNAs, organelles, and other entities inside it? What is the smallest element of "life"?

QUESTION 22:

"Could you use viruses as a supercomputer to explore the information space of cell entry or other biological mechanisms, just like we use AI for drug discovery?"

Paul Turner: (bio page 201)

Absolutely. This reminds me a bit of what we are trying to do with phage therapy. If you have a phage candidate that you want to use to target a bacterial pathogen, is it a problem if different people have that bacterial pathogen with a different genotype? Is my phage going to work the same in them as it works in you?

We are trying to use our huge experiments that look at different phages and different genotypes of the target pathogen in order to create a big grid of infection success (or not) for particular phages. This is accomplished by creating a more variable phage population to examine the possibilities and sort through all the data, especially at the genomic level. You want to make sense of it and perhaps take an AI approach, where you would train the system to understand the rules so that if I handed you a novel phage genome or a novel target pathogen genotype, your set of rules could determine whether it's a good phage match for that bacterial genotype, or a bacterial phenotype that will escape all the phages that we might throw at it.

I know you didn't exactly ask me that question, but harnessing phages or viruses as tools or as variability to look at outcomes is definitely happening in biotechnology. It's not a crazy idea at all. In fact, I think it's a great idea. It's more in the realm of what people call phage display, where they have harnessed phages for some period of time to create variable proteins and other things that could be tested through experiments to determine what binds to what, and to understand biology in some depth through a system that is itself variable, or that could be made highly variable to test different possibilities.[75]

Gareth Brady: (bio page 221)

In a sense, that's precisely how we use them because we are using them as sort of microcosms of evolution. RNA viruses evolve very rapidly because their genomes don't copy with the same fidelity as DNA viruses, like poxviruses. They evolve tremendously fast, so in a sense, the end product of the evolution of a virus inside of a particular host is like that experiment running.

With molluscum contagiosum virus, the lesions that form on the skin are completely chock-full of viruses and can last for years.[76] It's very unusual, because most viruses that last for years go into latency or dormancy; they hide from the immune system by switching off gene expression, or they deposit themselves in our genomes and just hide.

The molluscum contagiosum virus stays there for up to four years and is always producing all of its proteins.[76] The lesions are never inflamed, and largely persist as if the immune system doesn't even know that they are there.

This system seems impossible, because we know that the skin is the vanguard of the immune system. The keratinocytes in the skin express all of the sensors that I was talking about. Some cells express some of the sensors, other cells express other sensors, and the skin expresses all of them.

It is absolutely primed to sense any virus coming in, particularly molluscum contagiosum. We just need to figure out how it's doing it so that we can know exactly what bits in the pathway to go after in order to turn off human inflammation, for example, which arguably underlies all human disease, either directly or indirectly.

Marilyn J. Roossinck: (bio page 200)

I think that's actually quite possible and it's something I wanted to do. We all have lots of ideas that we can never actually bring to fruition. Some experiments are designed to try to get a virus to evolve to infect a different kind of host. The evolutionary capacity of viruses is very rarely appreciated.

I'll tell you of an experiment that I did in my lab. We were using a satellite RNA, which is a small parasitic RNA that's sometimes replicated by plant viruses. In this system, I could get two percent of the nucleotides to change within 10 days just by changing the environment.[77]

This is phenomenal, because that's the difference between our genes and the genes of chimpanzees. That's a pretty big leap, and it happened in 10 days. I tell you that to illustrate how rapidly viruses can evolve under the right selection pressure. They have huge mutational capacity, and I think this could be exploited more.

Richard Jacobs:

My semi-educated guess is that 1015 successful viral infections of living creatures have occurred over the past 3.8 billion years that life has existed on planet Earth. This number could be orders of magnitude higher, but is unlikely to be lower than 1012 at a minimum.

Amongst trillions or more successful infections and 1018 or more attempts at infection, even by modern synthesis/neo-Darwinian standards, there have been ample attempts, successes and failures over time for even random action to result in a nearly complete exploration of all the viable ways for a cell to be entered by a virus.

I don't know how many cataloged methods of viral entry have been discovered and documented. Data from countless scientific papers appear to demonstrate that a select few methods have been majorly successful, showing the following: (1) randomness may be more successful at engineering than humans are or ever will be, by not only finding a way, but finding it so reliably that a small subset of ways have become the predominant method by which viruses enter cells; and/or (2) the promiscuous nature of viruses to seek and infect hosts has allowed them to explore the most possible methods of cell attachment and entry.

Successful viral cell entry methods have out-competed unsuccessful methods and have trillions of successful infections to prove their efficacy. Whether the out-competing comes from deliberate trial and error or informed choice from viruses that are alive and deliberately adaptive, or whether it comes from random events acted on by selection pressure is immaterial.

By harnessing viruses in the trillions and exposing them to potential host cells under controlled conditions, the entire possible information space of cell entry by viral mechanism could be rapidly explored, much in the way that supercomputers with massive computational capacity are used to explore potential drug candidates for various biochemical processes.

A potential experiment might involve deliberately attempting to infect a collection of bacteria with a phage that doesn't normally infect that type of bacteria, then changing the pH, temperature, pressure, availability of nutrients, and upregulating or downregulating membrane receptors and other factors to see if viable cell entry methods arise.

Similar experiments could be conducted by using a cell that is already infected with a known virus, and changing that cell's local environment and other factors to slow or hasten various viral actions.

Another experiment would be to expose a known virus to previously infected cells that have some form of immunity in order to observe whether the virus is able to find an alternate method of cell entry. Scientists already know that this happens to people who have been previously infected by dengue.[78] This might be called biological supercomputing or active biological computation.

QUESTION 23:

"Do viruses contribute to an organism's immune system or fitness? Could our virome be a critical part of our immunity?"

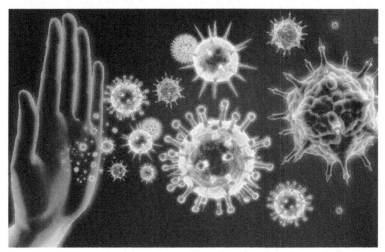

William B. Miller, Jr.: (bio page 206)

There are viruses that seem to be purely pathogenic to us. Smallpox would be one such example. We don't know one good thing about smallpox, but the vast majority of viruses are probably neutral and don't affect us one way or another.

However, there is another group that should be labelled as 'pathobionts'. The best way to explain that is by referencing a bacterium, Clostridium difficile. This bacterium, C. difficile, is a constituent of every normal gut microbiome, and it performs useful metabolic functions which help our gut lining and form part of our adaptive immune system.

However, if you get a serious infection that upsets the gut microbial ecology or take a series of antibiotics that selectively depletes some key bacterial players, you risk developing an overt C. difficile infection. Under certain circumstances, it becomes a substantial generator of pathology; these are very tough infections to eradicate.[79] A large proportion of our microbiome, whether bacteria, archaea, or viruses, are capable of having either effect.

In the end, it is best to look at cellular life as a form of symbiotic life among all the types of cells that are part of it and a vast viral compendium. It is a mistake to think of microbes and viruses as hangers-on, because they are very much a part of us as our own personal eukaryotic cells.

This is a tough concept for people, and my medical doctor friends don't like it. It makes people philosophically uncomfortable to think that they are, by definition, a living consortium. This idea intersects with free will and a lot of other issues that we are not going to go into here, but that is the way we need to look upon ourselves and that's how the virome needs to be explored.

Fortunately, this conversion of our thinking is active and ongoing. There are some great virologists, such as Luis Villarreal and a physician by the name of Frank Ryan who has written a lot about this duality concept under the term 'viral symbiogenesis'. Frank Ryan highlights it in his fine book, Virusphere.

It's not just about viruses as symbiotes with respect to our living health; it's about understanding that we are where we are on an evolutionary basis as a product of viral-cellular collaborative living principle. We can't have evolution without viruses being a critical intercessory between the cellular domains.[80] We exist as we do because viruses are there to provide one form of linkage. That's what viral symbiogenesis is all about.

Robert Siegel: (bio page 227)

First of all, one can make an argument that viruses actually gave rise to our immune system.[81] Viruses have two phases of existence: an inert phase outside the cell, where they don't move or metabolize or do anything like that, and a dynamic phase, where once they enter a cell, their genetic machinery starts taking over every aspect of the cell's behavior.

The immune system is also divided in two: there is the humoral branch which produces antibodies that inactivate extracellular or inert phase viruses, and there is also cell-mediated or T-cell immunity, which is able to detect and kill virally-infected cells. Right now, we don't have a good way of eliminating the virus from those cells, so we sacrifice infected cells in order to prevent them from making more virus that could be passed on to other cells.

In this sense, viruses have very much affected our immune system, which is helpful in all sorts of ways, particularly for long-lived creatures that have to deal with lots of different pathogens and environmental assaults. We cannot respond by evolving really fast, so instead we have developed powerful adaptive defenses.

Lily Wang: (bio page 218)

I think so. When our ancestors were infected with a certain agent or virus, it would sometimes be stored inside the genome. In fact, about 10 to 15 percent of the whole human genome is infected by ancient endogenous retroviruses, and there might be viruses in addition to retroviruses. I think that the storage of this information may help human beings fight against those species of viruses with a known mechanism.

Joseph R. Masci: (bio page 216)

In terms of how viruses can enhance or protect human cells, I am not sure if there are data on that. It is of course being looked at in the laboratory, but I don't know a natural example of how introducing a viral agent could enhance anything about human cells. As far as I know, there are only theoretical, potential advantages.

Gareth Brady: (bio page 221)

I think they indirectly beneficially contribute to it. The process is like the arms race in a cold war. The immune system has evolved to be complex because viruses have gotten better and better at evading these pathways. For example, Drosophila is a fruit fly that is known as a very amenable system for genetic study. Drosophila has a reduced stash of these sensors and systems.

If you look at a much more complex organism like humans, there is a massively expanded set of pathways and sensors. I believe this is because the viruses that infect humans have been targeting these pathways for so long that humans have been forced to evolve multiple sensors to detect them.

At the same time, these viruses are evolving more and more inhibitors to target these pathways. As viruses get better at avoiding these pathways, vertebrates like humans have to develop new ways of stopping them from slipping under the radar.

The virus is like a burglar in a diamond shop: if the burglar gets good at avoiding certain tripwires of the immune system, then you have to design more nefarious ways of triggering, injuring, or destroying the burglar before they steal the diamond.

In a sense, the virus is driving our evolution in a way that allows us to become better at sensing pathogens. Some of these sensing pathways have the fringe benefit of detecting bacterial and cellular damage. For example, a protein called cGAS was discovered in 2016 by James Chen. This protein doesn't specifically detect viral DNA, but it almost certainly evolved to detect DNA viruses.[82]

In knockout experiments in mice, we can see that they very much have a problem detecting DNA viruses. Since cGAS evolved to detect DNA in a non-specific way, it is just looking for decompartmentalized DNA. All of our DNA is in the nucleus and some of it is in the mitochondria, but none of it is in the cytoplasm. We have a system that detects DNA in the cytoplasm as viral DNA.

As it happens, there are also bacteria that can infect cells (e.g., Mycobacterium tuberculosis); once inside cells, the bacteria are detected by cGAS.[83] Furthermore, when a cell nucleus gets damaged and DNA leeches out, cGAS triggers and accelerates the process of wound healing.[84]

Viruses have actually benefitted us by forcing us to derive these new systems for detecting not only their presence, but also the presence of lots of other things. It's massively speeding up evolution in that respect by forcing us to adapt constantly to the outside infectious environment.

Richard Jacobs:

Let's take the case of CTX prophages infecting Vibrio cholerae and transmitting essential viral genes to make this type of bacteria pathogenic to humans. V. cholerae inside of a person's gut is likely to come under immune system attack, and compete with other gut bacteria for nutrients.

To maintain their ability to live, multiply, and proliferate, they are at a competitive disadvantage before they obtain the CTX prophage genetic material.[85] In this way, the phage is acting as an external immune system to the V. cholerae, improving its fitness by "gifting it" some of its genetic material (or by V. cholerae co-opting this part of CTX prophage viral action).

If we consider the hundreds of bacterial species in our microbiome, it is highly likely that our phageome acts as a usable library of genetic ability. Although a large part of phage action is pathogenic and detrimental to the

154

bacteria they infect, there is a significant number of instances of infected bacteria forming a mutualistic relationship with their prophages and benefitting from their association with them (as in the case of V. cholerae).

In the same way that our microbiome benefits and is vital to our health, our virome is only minorly pathogenic,[86] at best. Given the fact that we are continually inhabited by many different types of viruses[87] and that eight to 10 percent of our genetic material has been obtained from viruses, it makes sense to assert that if our virome was substantially pathogenic, then our health would be negatively impacted on a continual basis, and billions of people would not experience healthy living that can persist for decades before illness strikes.

Our own DNA has benefitted tremendously from the endogenization of viral genes, including placenta-forming syncytin proteins, which are vital to the successful procreation and perpetuation of placental mammals (especially humans, of which you the reader, are a member in good standing). This means that on dozens, hundreds, or even thousands of occasions, our bodies have withdrawn books from their local virus libraries and learned (endogenized) a thing or two.

Frank Ryan, author of Virusphere, Virolution, and other books, talks about how viral infection even shapes and guides evolution itself. From facilitating immunity to providing positively adaptive genes, viruses play a major role in our lives and existence.

QUESTION 24:

"Do you believe that our somatic cells communicate directly with our microbiome, including bacteria, viruses, fungi, etc., or is communication siloed between viruses and bacteria only?"

Nils Walter: (bio page 224)

Scientists have been doing biology at the molecular level for a little over 100 years, which isn't a very long time, especially in the historical and geological context. I think we are finally in a position to understand all of the components in a human cell. There are cell atlas projects that basically catalog all of the components of the cell and include information about when they are made, when they are expressed, and under what conditions these things happen.

From the Human Genome Project, we know all of the sequences and can therefore derive information about RNAs and proteins. What we don't fully understand yet is how all of these things are interacting with one another. Currently, we are trying to understand all of the pathways in the cell, but we already know many of the basic pathways that make metabolites, vitamins, and the building blocks of life, because they were studied in the 1960s, 70s, and 80s.

Only recently have we learned that there is a whole set of RNA metabolic pathways.[88] Most of the components we now know, but we don't know how they interact with one another. We are in a position where we can actually understand what is called the 'interactome,' or the entirety of all these interactions between the catalogued components of a human cell.

To our surprise (or maybe not to our surprise), they all interact with one another within the cell. Everything can bind to a lot of other things, sometimes very briefly, sometimes for long periods of time. If something binds for a long period of time, it might induce a new process or pathway, or it might be sequestered away from a different process. We are starting to understand how sequestration can occur, and how the interactome works.

On top of that, there are 60 to 100 trillion cells in our body (depending on whether you include bacteria),[89] which all interact with each other and exchange metabolites and pieces of genetic information. The interactome across the 60 trillion cells is even vaster than what is happening inside a single cell. Now, factor into that the exchange of information that occurs not just with the food that we eat, but the nucleic acids that we take up through our food. Most of that gets degraded, but some of it passes through. Certainly, some things that we take up directly into our bloodstream or into our saliva might get into our bodies in some way. I imagine that 99 percent of the time, it's degraded, totally harmless, and even used as new building blocks.

In other cases, however, it could be an intact virus interacting, which leading to an epigenetic mark that changes the expression pattern in a subset of cells, or changes the likelihood that we get cardiovascular disease. If two things interact with each other, there is one interaction. If three things interact with one another, there are three different paired interactions, and it goes up from there.

Now imagine 100 trillion cells in our body, the environment in which we are enclosed, and millions of different molecules in each cell which interact with one another in ways that we have not fully understood. Ultimately, that is what life is.

Of course, with the catalog in hand, we can start tracing the interactions inside the cell, among cells in the body, across organisms, and even across the world, if you wish. That's a lot to study and it will take some more time to learn. This is a century where we are making a lot of progress.

Güenther Witzany: (bio page 210)

I can't imagine a coordinated behavior of cellular organisms without communication. How should this function? For example, in the liver or the stomach, the cells have to coordinate in order to carry out a tissue-specific behavior, and this coordination depends on good functioning communication. If the communication is disturbed, then the result will be a disease of the organism. If the communication functions, the organism will grow and prosper.

It's a constant communication process which is never sleeping; it may be more silent or less active at times, but there is constant communication between all cells of an organism.

Lily Wang: (bio page 218)

We have established that there are several species of bacteria in our gut that are beneficial for the human being.[90] These bacteria can be deemed part of the human body, because if the balance of the bacteria in the gut gets disrupted, people will get ill or develop diarrhea, etc. I believe the virus might play a role in that.

Joseph R. Masci: (bio page 216)

I think they interact. The fecal microbiome has become important because there are tons of bacteria in our large bowel, and they are clearly serving a purpose. One major purpose is to suppress the overgrowth of virulent bacteria. Bacteria have a suppressive effect when everything is working well on bacteria that cause disease.

One of the theories of inflammatory bowel disease is that it is just an imbalance of the bacterial population;[91] where viruses fit into that is not 100 percent clear to me. We haven't discussed how viruses can alter the products that bacteria put out or alter the genetics of bacteria, but viruses may actually interfere with or enhance the ability of bacteria to suppress bad bacteria.

There is a reason we have so many bacteria in our large bowel; it's not just because there is stasis and overgrowth, but because they serve a genuinely important purpose.[92] As a side note, this has become more of a concern with the overuse of antibacterial drugs in young children, in particular where antibiotics can alter the intestinal microflora for a surprisingly long period of time.[93]

This might be placing children at risk for other illnesses yet to be identified, but that can arise from a distorted intestinal microflora. I think everyone is recognizing now that there is a lot of importance to the intestinal microflora in terms of how it can lead to good and bad health.

James Shapiro: (bio page 202)

We know the microbiome is directly communicating with our cells because we know that changes in our microbiome can influence our health and our mood. Microbes are not just hangers-on; they can influence the way we think, the way we feel, what we can digest, and how active or passive we are.[94] They influence us in many ways and that's a whole active field of research these days.

Richard Jacobs:

I've asked many podcast guests, "Who eats first, us or our microbiome?" The answer may be that our somatic cells "eat first," and the undigestible remains are passed along to our gut bacteria, but this doesn't answer the question of why people have various microbiomes in hundreds of places in and around their body — places that expose the food and liquid we consume to microbes before we digest it, such as in our saliva.

If you think about it, food that we eat and liquids we drink (except maybe processed foods that not only have nutrients and bacteria stripped from them, but preservatives that keep out certain microbes and keep the food "fresh") have already been acted upon by bacteria and other microbes. Cheese, kombucha, sauerkraut, kimchi, coffee, chocolate, bread, and thousands of other foods are actively fermented or changed by microbial action before we eat them, AND after we eat them.[95]

In addition, I've read in the scientific literature that thyroid hormone is made bioavailable by our microbiome (T4 to T3 converted), and many of the B vitamins are also created by our microbiome.[96] Eighty percent (80%) of all serotonin in the body is made in our guts by bacteria.[97]

In order for our bodies to function properly, our somatic cells MUST be in communication with our microbiomes and vice versa. There's a whole lot of trading going on. Even if some of the communication is under the guise of combat and defense of self (immune reactions), the trading of molecules back and forth must be well understood and governed by rules.

Although this is an unknown and unexplored concept, what of "bacterial economics"? How many glucose molecules does a certain bacterium "expect" in return for producing molecules of serotonin or one of the B vitamins? Is it a one-to-one ratio? Are other trades a three-to-one ratio, or even a one-to-one-hundred ratio? Who decides how much each molecule is "worth," based on the effort needed to create it and what it is worth in trade for another molecule?

In this book and in the literature, there are examples of phages interacting with bacteria (this happens constantly), viruses interacting with our somatic cells, and our somatic cells interacting with some of the players of our microbiome (bacterial examples most common), but there are likely interactions amongst all the creatures inside of us.

Imagine shrinking yourself down to the size of one of our gut cells; what would you experience, and what would your sensory apparatus be like? Imagine shrinking yourself down to the size of a bacterium, where you're still 100 times larger than most viruses, but 100 times smaller than many somatic cells; what would you experience in this size range?

Now imagine shrinking yourself down to the size of a common virus, which is 50 to 100 nanometers. There are countless virions (both phage and human virus type), extracellular vesicles, gigantic bacteria and even more gigantic somatic cells, large protists and fungi, small molecules floating everywhere, and more.

I think that there is a degree of siloing of communication, but also massively interlinked communication across all cellular and viral domains.

QUESTION 25:

"Virions are mostly or completely non-motile/passive and tiny in size (50-100 nm) in relation to a macroscopic host such as a human, yet they infect in tremendous numbers 10(superscript 12) virus particles infecting?); Could brownian motion and chance be responsible for countless 10(superscript 30?) Successful viral infections of countless target cells and creatures over the past 3.5 billion+ years? Or... is there a method of sensing and guidance that viruses employ in their virion stage when moving through a host to target specific cell types?"

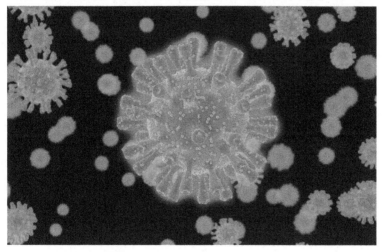

Paul Turner: (bio page 201)

To me, this is something that is staggeringly difficult for the human mind to comprehend. It is somewhat equivalent to the stars in the universe, or the size of the universe; we can't really contemplate it because we don't have a good grasp of the scale of it. The problem for us is that we have a difficult time thinking about what it is to navigate the world as the size of a virus.

I tend to think of it in terms of the enormity and efficiency of viral replication. Most viruses make progeny very efficiently, and it's a numbers game; many of them are very unsuccessful and a handful are successful.

I'll extend this analogy. A jellyfish releases millions of juvenile jellyfish, each requiring a key developmental stage during which they adhere to a surface; how does that happen, and how are they successful in the vastness of the ocean where they don't have much power over where they'll go? They are moved around by the physical forces in the water, so how are they successful?

To me, it seems like they are successful because they produce so much progeny that the lineage goes forward even though very few individuals are successful in making it to the adult stage. For a virus, the adult stage would mean infecting a cell, replicating, and making progeny. If that's inefficient over time, there are enough virus particles of any one type to overcome the problem.

Marilyn J. Roossinck: (bio page 200)

Every virus has a different mechanism. It's difficult not to generalize, but you can't generalize too much with viruses because they are all different and have found different ways to get into their hosts. It's not that there is a big sea of viruses all around us and once in a while they bump into the host cell; instead, they have active mechanisms. For example, most plant viruses are transmitted by insects, and can be transmitted by fungi and other things.

A plant has a cell wall, so for a virus to get into the plant cell, it has to make a hole in that cell wall. Insects do this very efficiently, so viruses are moved from one plant to another by insects. Viruses control that process to some extent, because when they are infecting a plant, they will change the volatile composition of the plant, and the plant will start making qualities that will attract insects to the plant.[98]

Once an insect has landed and starts to feed on the plant, it can take up the virus. The virus then instructs the plant to make anti-feeding compounds to encourage the insect to move off to a new plant.[98] The virus has manipulated this whole system to help itself move around. These relationships are very old and often very specific. There are some viruses that are only transmitted by a few species of aphids, for example, and there are others that are transmitted by hundreds of species of aphids. This is just one example.

Jeremy Barr: (bio page 217)

This is such a cool thought experiment. I love this scale. As big as the divide is between the size of the virus and its target, the number of viruses is so many orders of magnitude greater than that difference in scale. It's estimated that there are 1031 phages on the planet, which is a number that no one can even begin to comprehend. The scale is just beyond our comprehension.

If we take the perspective from within the gut; this is a hugely complicated and diverse ecosystem, so how can a single virus find a bacterium within it? I'd say that it can't; the chances of that are infinitesimally small, so it would probably never happen. The only way that it can happen is by having these ridiculously large numbers of virions. How many viral infections are successful? I don't think we'll ever know the answer to that, but it will be incredibly small percentage. If there are 1031 phages, at least 90 percent of them have failed to infect, or are just sitting in the environment hoping and waiting.

Viruses can overcome this problem of scale and finding a host across both space and time by flooding the environment with viral numbers that we can't comprehend. When you flood a system with that many viruses, encounters that were borderline impossible are almost absolutely going to happen. This is how natural diffusion and persistence in the environment is going to drive these chance encounters, which is what will drive more and more viruses to persist.

As a post-doctoral student, I conducted research alongside Forest Rohwer on how a virus finds and encounters a host. We looked at the mucosal surface in the gut, where there are cells that produce a mucus layer that's critical in the support of the microbiome. We found that certain phages have specific proteins on their heads which allow them to stick to the mucus.[99] The attachment is weak, so after sticking for a little while, the phages detach and keep diffusing in an on-and-off motion.

Initially, we found that by sticking to this mucus layer, the phages protect the mucous cells from bacterial infection.[99] Next, we started looking into how these phages diffuse. We watched them in real time under a microscope and tracked them in water, where they diffused normally. In mucus, however, they diffused sub-diffusively, which refers to really complicated mathematics behind diffusion. Simply put, we were able to show that the phages would diffuse normally and then stick to mucus for an undefined period of time before diffusing again with an on-and-off motion, and we considered it to be a search strategy.[99]

Gareth Brady: (bio page 221)

This question doesn't really take into account how fast microscopic things move in fluid; they move dramatically fast. It's driven by tropism for the entry receptor. For example, if HIV enters the bloodstream through fragile tissue or broken skin near the genitals, the virus will have to go quite the distance before it comes across a CD4-positive or CCR5-positive cell.

It is looking for both of those receptors, and they both happen to be expressed on CD4-positive T cells. They are floating around in this fluid all the time at a very fast speed, but so are the T cells. When they come into close proximity, the interaction between the virus receptor and the CD4 and CCR5 molecule is electrostatic, like a magnet.[100]

Electrostatic interactions decrease with the square of the distance between the two things interacting. However, there is still a magnet-like attraction between those two things because they want to find each other from a molecular point of view, so it doesn't take too much time.

For example, in the bloodstream, the virus is going to be moving throughout the whole body as the blood is pumped through the heart and percolates through the tissues. It's not going to be too long before it finds those cells, and once it does, it will start to replicate and produce 106 versions of virus per ~50,000. This creates a huge viral load, so at that point, it won't have any trouble finding the cells.

There is a lag period between entry and finding the target cell, but it's not really as long as you think because of the nature of how fast microscopic things move in fluid. It's not the virus itself that is moving; it's fluid motion that's pushing things around, and you can clearly see this under a microscope.[101] It's not motility per se, but fluid motion.

Forest Rohwer: (bio page 196)

In the end, it really is just a numbers game. There are just so many possibilities for them to be able to do these cool things that I would say you can equate the success of viruses to the fact that there are just so many of them, and so many possible ways of making a living.

Richard Jacobs:

Imagine yourself the size of an ant in an Olympic-sized swimming pool. If you had no means of moving yourself about, the likelihood of you finding the right 20-80 nm receptor on the right cell target in that vast expanse would be infinitesimally small. Of course, there would be trillions of you or more; however, not only would you have to find your target receptor, but you'd have to align your body to bind to that receptor properly.

Certain shapes of virions are more amenable to harnessing random motion to bind to a receptor. Spherical coronaviruses, for instance, have radially distributed spike proteins that are more likely to contact the right receptor at the right angle than rod-like virions. [102]

Some virions have whip-like hairs that appear to feel their way over the surface membrane of cells; others "dock" with a receptor, then change their structure to fuse to a cell's membrane. Others appear to be more passive, or deliberately engineered to change shape to continue a sophisticated entry pattern. Still others maintain their outer coatings and shape until they enter into a cell's cytoplasm inside of a vesicle, then exit that vesicle, open up, and deliver their payload.

At the very least, virions are passive only up to a certain point, and are exquisitely adapted to sensing the right binding situation with the right receptor, and then actively changing their structure and shape to finish entering the target cell.

Once inside a cell, virions appear to become extremely active, such as by using the cell's machinery and sensory apparatuses to monitor the inner cell and extracellular environment. I don't think that passive, non-motile virions are an accurate picture of viral action, and I don't think that scientists know for sure what the real picture is for all known viruses.

QUESTION 26:

"Which came first, viruses or cellular life forms, and why?"

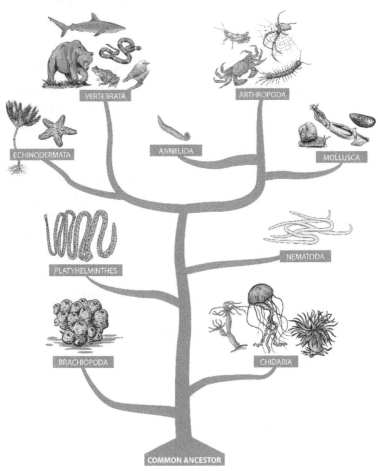

Denis Noble: (bio page 198)

There is no way that viruses could have come before cellular life forms because they can't replicate without a cellular life form. Having said that, we come to the question of, "What is a virus?" Could extracellular vesicles have themselves become viruses? Well, of course, they'd have to be cells in order to produce the extracellular vesicle, so that doesn't answer by itself the question of which came first. I can't see how anything that cannot self-replicate could have come first.

How many properties of life would have had to be present is an interesting question as well. There would have had to be some way of enclosing a metabolic pathway or network of interactions between various chemicals sufficiently restrained to ensure that once the reaction gets going, it doesn't immediately disperse in the primeval sea in which it finds itself.

Certainly, whether cell membranes were there in the first self-reproducing metabolic pathways or networks of interactions is a very good question. Lee Cronin at Glasgow University has done some fascinating work on this topic.

Specially, he's looking at the way in which iterative modification of polypeptides (the precursors of proteins) could have developed into a self-maintaining system.[103] I'm inclined to think that's the better way to look for the origin of life than to imagine that life started with viruses.

Marilyn J. Roossinck: (bio page 200)

Personally, I think viruses came first, but that's just speculation. We don't have any real evidence, but there is a whole hypothesis about the early world which would imply that viruses came first. Without a time machine, we really can't tell. It's not a question that is easy to answer, but there are people with pretty strong opinions.

Lily Wang: (bio page 218)

A virus needs a cell mechanism to replicate, so I'm pretty sure cellular life forms came first. The virus might have originated from some kind of a plasmid, and information could have been transmitted between cellular life. Later, the plasmid-like materials could have gained the ability to produce the structural proteins and assemble themselves into a virus-like form, which would make it easier for them to transmit between the cells. I think that the original viruses were just simple plasmids.

Joseph R. Masci: (bio page 216)

I don't know if there is an agreed-upon answer to this question. I think it's tempting to think that viruses came after cells because cells are infected by viruses, not the other way around. However, whether they developed before cells and were able to survive without an intracellular existence in some other way is not clear to us.

I think there are examples of how a virus can live outside of cells in blood plasma, at least for a period of time. It could be argued that there was a similar environment that allowed viruses (if you want to call them that) to exist before they became parasites of cells (since there weren't any cells that allowed viruses to exist in terms of replicating RNA strands).

Since these organisms don't leave fossils, I think it's going to remain very speculative, but I think there is every reason to support either hypothesis—that cells came first and viruses started hitching a ride, or that viruses came first. There is a third hypothesis that neither came first, but that each evolved separately and they eventually met up with each other.[104] Fascinating.

Güenther Witzany: (bio page 210)

Currently, we have strong reason to believe that there were RNAs without cellular organisms.[105] An abundant mass of RNAs in the ocean cooperated and formed networks of various forms. We have one virus which has nothing else as its RNA sequence. These are the viroid of plants or plant viruses. The RNA sequence is able to interact with the host organism or within a group of RNA networks. Later, cells started. How did cells start? There had to be a membrane, because a single cell without a membrane will not survive.

Current investigations show that viruses produce some protein membranes, and it is thinkable that in some environmental circumstances (e.g., soil, deep sea) there is compartmentalization. If there are compartmentalized regions where RNA groups are isolated, and if they have protection by proteins that are used by viruses, then this could be the start of the first cells.

I don't think that viruses came before cellular organisms, but the cellular evolution of the first cellular organisms without viruses is unthinkable. In a parallel way, they found a way to create cellular life together, with viruses based on RNA networks.

Richard Jacobs:

This is even more complicated than the chicken and egg question. It doesn't appear that chicken eggs could've existed on their own before chickens appeared, while viruses—especially because they contain so much novel genetic material and novel genes that do not appear to exist in any creature—may have appeared before cellular life... or vice versa.

In comparing the biogenesis of extracellular vesicles, the action of sperm (similar to viral action when considering the events needed for successful conception), the biogenesis of extracellular vesicle cargo that contains fully-enclosed RNA in viral-like capsids, and the similarity to bacterial plasmids, eukaryotic tunneling nanotubes, and other cellular tools and abilities, viruses may have required the existence of cellular life first.

Viruses may be renegade or rogue cellular creations/tools, that have unintended targets.

"What role do viruses play in evolution, adaptation, and speciation?"

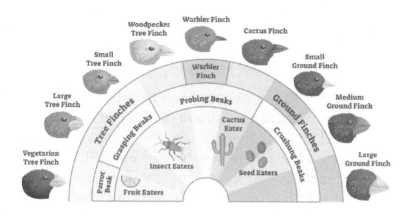

Paul Offit: (bio page 211)

I think that they've certainly changed the course of human history; smallpox killed over 500 million people, and the Spanish flu pandemic of 1917 and 1918 killed about 100 million people. Endogenous retroviruses and bacteriophages are at a level of 1030. You have 10 times more bacteria on the surface of your body than you have cells in your body, which determine whether you are likely to get diabetes, obesity, asthma, or various allergies. I think that the viruses that are so much an integral part of the bacteria are no doubt a part of that.

Nathalie Gontier: (bio page 219)

I think viruses are major players in evolution insofar as they provide an opportunity for organisms to evolve. If a species is attacked by a virus, there are only so many routes it can take; it can die or learn to co-exist with it. We know that organisms are able to change the genes of a virus in order for the virus to become less harmful, and that a virus can somehow find a space where it can integrate and become part of the genome. I think that viruses are major players in biodiversity, in a good and bad way.

With the COVID-19 virus, there have been 19 million infected individuals, and almost one million individuals have died. If you look at all these bioconservation programs, there are species that are infected in much fewer numbers. A virus can facilitate or induce extinction, and it can also help with adaptation and perhaps speciation insofar as both species evolve as the result of an interaction between them.

Marilyn J. Roossinck: (bio page 200)

Eight percent of our genome is comprised of retroviruses, so they've been with us since the beginning of human evolution. Many of those are shared by other primates, and there are some that are even shared with coelacanths, which are considered living fossils.[106]

Most obviously, retroviruses have a played a role in evolution because they are in our genomes and there are retroviral genes that are critical for our survival. For example, the evolution of the placenta in mammals required a retrovirus. We don't have fossil records of viruses themselves because they are too small and are not maintained in stone, but there is a paleo-virology field within which people research viruses that are integrated into genomes; this is the equivalent of the genetic fossil record.

We are not really separate from our viruses; our viruses are a part of us, it's just that they have to get into a host. What is a human being? We are not just human cells; we have bacteria, viruses, and fungi, which comprise the whole biome as a single entity.

Eugene V. Koonin: (bio page 194)

Viruses play a huge role in evolution, speciation, and adaptation at different levels. For example, by infecting hosts that have defense systems that sufficiently protect the host, viruses can affect the phases of different populations, and even of different species. They may accordingly affect the composition of any ecosystem if a resistant strain of species develops efficient defense systems, and survives the founder effect of other genes or other properties that the population shares. The virus hitchhikes on their defense system and survives.

Viruses affect evolution, but there are other types of properties that hosts acquire, and they adopt viral genes for their own function. A non-textbook example is the receptors for a fetus on the mammalian placenta. These are proteins that have been acquired by the ancestor of the placental mammals from retroviruses, and in the retroviruses are proteins which form the virus antidote. If these proteins didn't exist, then we would not be having these conversations because we would not exist in any form close to our present form.

There are quite a few such examples. After all, we have to keep in mind that a very significant fraction of the genome in complex organisms such as mammals, plants, and invertebrates consists of remnants of viral genomes. There are so many of these theories, and clearly, they are used for various types of calculations. Viruses are an indispensable component of the evolution of all organisms, and in particular, of complex organisms such as ourselves.

James Shapiro: (bio page 202)

They play a number of roles. One role that is very obvious and wasn't expected when it was first discovered is that they serve as vectors to carry information from one kind of organism to another.[107] Viruses can pick up information in one kind of cell and take it to another kind of cell. Bacteria can live in the same amoeba and pick up sequences from giant viruses, and the giant viruses can pick up sequences from the bacteria.[108] Some of these bacteria can infect mammals, for example, and they might carry DNA from the virus to the mammal; this is called the amoebal evolutionary melting pot.[109]

I like to refer to viruses as the "research and development" sector of evolution, because there's a lot about viruses that is quite mysterious. We don't know where much of the viral information comes from, and close to 99 percent of all the sequences which are found only once in the databases are in viruses.[110] Those are called ORFans because they are alone, and ORF means "open reading frame." Those sequences are a source of information for other organisms in evolution, and we see a lot of virus-to-host transfer in the making of new kinds of proteins. We don't know how those sequences arise, and in many cases, we don't know what they do or what they code for. Viruses are springs of innovation, which is why I call them the research and development sector of evolution.

Richard Jacobs:

I hypothesize that all living organisms are continually adapting their environments within their biological capacity to do so. From time to time, such as when a meteorite impacted the Yucatan Peninsula 65 million years ago, causing the apparent extinction of many dinosaur species, conditions may be so extreme as to cause speciation via forced adaptation that is outside the purview of epigenetic marks, changes in gene expression, and other adaptations.

Eight to 10 percent of our own DNA is comprised of viral genes that must have endogenized over the past 200,000 to 20 million years. A non-endogenized viral infection may be completely or partially lethal, at times latent and commensal, and in some cases, mutualistic in that it spares only certain infected individuals who are "carriers," and actively attacks competitors that would mate with others, ensuring the passage of carrier-type individuals over time until they predominate in a species.

Myxoma virus was introduced to cull rabbit populations in Australia and led to the death of 99.8 percent of the rabbits in Australia, sparing only a tiny percentage that were somewhat immune.[111] Over time, the carrier rabbits outbred the other rabbits, to the point that the entire Australian rabbit population is now composed of members immune to Myxoma virus.[111]

Although this has not caused speciation, a series of events such as this that occurs over time, usually naturally caused, could lead to a continual adaptation and change of a species until it reaches the point where it would be unable to reproduce with previous versions of this species, slowly creating a new species.

Over the course of our lives, our microbiomes change, our cells senesce, we become epigenetically marked, and we change, although we don't change enough to become a new species... but over thousands of generations, the compounding action of infecting viruses and changing environmental conditions could steer a species towards becoming a new species.

Viruses are a powerful force in forced adaptation, in shaping our immune systems, in causing speciation, and also co-evolving with all life.[112]

QUESTION 28:

"Some viruses, like HIV, endogenize yet preserve their agency, while others endogenize and lose their ability to replicate. What dictates whether a virus will maintain its agency, and when might it lose it?"

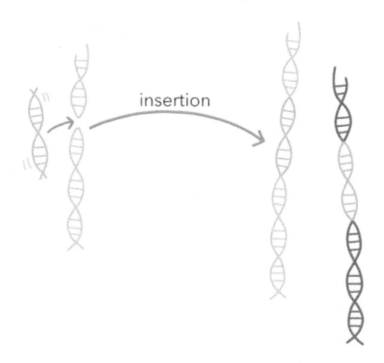

insertion

Nils Walter: (bio page 224)

I think they retain independent agency for a certain period of time, but it's kind of like a muscle; if you don't use a muscle, it will lose protein and become very thin. It is the same with the ability of a virus to replicate. If it inserts into a genome and the cell itself divides every so often and replicates the retroviral DNA with its own DNA, then the virus itself obviously doesn't need its own replication machinery. The cell will do everything it can to silence the virus, and it may put epigenetic marks on it so that the virus cannot replicate, in which case the virus would be silenced and unable to do anything.

The virus piggybacks on the replication machinery of the cell itself, and its genome gets replicated. However, because the virus never uses its own replication machinery, over time, there is a probability of about one in a billion per nucleotide that a mutation will occur. That is, in each replication of a human cell, there is roughly one error made for the entire genome. Additionally, mutations or other changes can occur spontaneously through unrepaired damages.

Occasionally, a mutation will become inserted where the virus is found in the genome. Over time (it may take generations), the information of the virus to replicate itself will be lost.

After hundreds of generations of cells and humans, that retrovirus may become something that can no longer replicate but still has similarity to a retroviral genome. Over the lifetime of just a single person who is infected with HIV, it's still likely that at least some of these viruses will stick around in their original wild-type sequence, eventually come back to life, and have a certain probability of killing the person unless the person receives drug therapy.

Michael Betts: (bio page 223)

Human Endogenous RetroViruses (HERVs) are distantly related to HIV and have been with us for millions of years. At some point, they were probably pathogenic and transmitted, but they do not seem to be anymore. They have deletions in parts of their genomes that are not full viruses anymore, but can clearly be recognized as being of viral origin.

It's actually thought that these HERVs and other related elements have allowed us to evolve to be where we are. In these kinds of situations, you would envision that these retroviruses were carrying a gene that they needed for themselves, and that also turned out to be beneficial to the host. We were better off having them and being infected by them, so they became a stable part of our genome. Maybe that is partially how humans have evolved to the current state.

James Shapiro: (bio page 202)

There are people who think that HIV is on the path to eventually becoming part of the human genome. When you first encounter a virus, it can be much more dangerous than it would be after some period of familiarity with it.

While some people use the terms 'sense of self' and 'independence,' I think viruses are just part of the whole biosphere, the living world. They go into cells and come out of cells and can communicate between cells, and they can sometimes affect other properties of cells. It's a little bit strange to talk about a virus as having an identity or a sense of self, as this is like saying that certain molecules in a cell have an identity or a sense of self.

Forest Rohwer: (bio page 196)

There is definitely proviral degradation, which is when viruses lose essential bits over time and essentially become trapped in the genome. What's really interesting is that a lot of them can actually use related viruses to become active again.[113] Even though it looks like the provirus is dead, we frequently find that the provirus has re-entered an active phase or virion and is sewing it using pieces and parts of other viruses. It is kind of like complementation between different viruses. The endogenous retroelements in humans seem to get around even though they are 'dead' in terms of not having all of the components it takes to make them, but they can still be activated by other viruses.[113]

Luis Villarreal: (bio page 203)

Think about the thought experiment we did with HIV running unrestrained in human populations. We would predict that you'd end up with a population of humans that were non-progressors, and that that population would no longer be compatible with the prior human populations. If you try to mate them or bring them together in a common social exchange, one becomes lethal for the other, exactly like the lysogenic phage.

We do expect to see this population phenomenon. We would say the identity, or the colonized version of the human, has acquired a new molecular identity that's associated with the capacity for harm to peoples that aren't a member of that same identity population. The group identity has changed and has acquired as a new viral feature. Now you can see why this process tends to be acquisitional, and why you tend to get more of these agents with evolution. With time, it builds towards complexity and towards a more complicated group identity.

Richard Jacobs:

From what I've observed, no one knows the answer to this question. What is known is that dozens — possibly hundreds — of viruses have endogenized themselves into the human genome (and the genomes of other animals and creatures) since Homo sapiens first arose, at least 200,000 years ago. Long-term endogenized viruses appear to have lost some of their genes, which allows them to act independently from their host. In humans, these viruses are called HERVs — Human Endogenous RetroViruses.

What is amazing is that endogenized viral genes appear to be NECESSARY for humans to exist. Take the case of syncytin, a viral protein that is responsible for the barrier between mother and child; it forms some of the placental membranes, allowing for placental mammals to exist. Without it, no placental mammals, and no humans!

I hypothesize that viruses retain some aspects of self and agency, even when they first endogenize into a host (like HIV, which endogenizes, but can still cause an infected cell to produce copies of itself and package them into capsids and make complete virions). I hypothesize that a virocell is created (credit to Patrick Forterre), and that this hybrid organism allows for many viruses to "know" that the infected cell "knows," and to have access to many of that cell's abilities and to its sensory apparatus (i.e., cellular senome).

If the endogenized virus doesn't kill the cell over time through co-adaptation and co-evolution, then it will drop some of its redundant equipment (genes) and retain only the components of itself necessary to enhance the continuation of the whole — of the merged identity of the virocell, which to an outside observer, merely becomes "the cell" over time.

Scientists don't consider human cells to be virocells in their current state, although eight to 10 percent or more of all human DNA is viral in origin. At what point and under what circumstances scientists should consider an infected cell to not only be a virocell, but a cell, is not fleshed out.

"Do viruses acquire epigenetic marks, and if so, what are the consequences of that?"

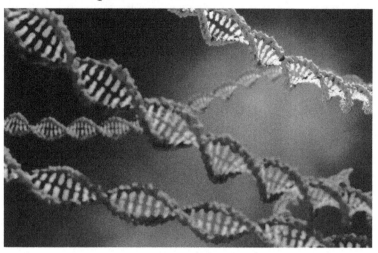

Richard Allen White III: (bio page 209)

Absolutely. Phage DNA is heavily modified. There are probably modifications in phage DNA that we've never seen before, because if it modifies its nucleic acid, that can't be chewed up by host nucleases. Papers are being published every month about a new phage that has methylation or some sort of modification of nucleic acid that doesn't allow it to be chewed up. I think that's absolutely fascinating. I think there is a tremendous amount of work that can still be done.

Paul Turner: (bio page 201)

It would not surprise me. I don't have great evidence or examples of that in my own research, but I'm going to describe something that's not completely unknown in viruses, meaning we have some evidence for it:

If a virus is replicating in cell type A and in cell type B, then as those virus particles exit, they could present the exact same genotype, yet have very different abilities in terms of infecting other cells of type A or type B, or even a cell of type C. In other words, there could be a maternal effect, if you will, from replicating in one cell type that does not alter the genetic material, but somehow alters the phenotype, performance, or ability to infect another cell and successfully undergo replication.

What causes that? Is it something that is attached to the virus's genetic material in some epigenetic way? Is it the proteins or lipids on the exterior of the virus particle that it obtained from being in cell type A vs. cell type B, and that affect how the virus replicates in a similar or different cell type?

I don't think we have a great answer for why this happens, but it's profoundly important because it's at the root of what's called 'emergence' or 'host shifts.' How is it that viruses exit one host site and successfully (or not) replicate in another host site? We need to figure out the rules and mechanisms behind these observations.

Forest Rohwer: (bio page 196)

They are kind of the original epigenetics. What we consider the main defense of viruses against the cell are restriction modification systems. These are two component systems where you have a modification and usually a methylation of that, which then protects against the restriction enzyme that enters with the virus. There are different methylation patterns depending on how viruses replicate. You definitely get epigenetic imprinting on viruses.

Richard Jacobs:

When considering animal cells (and probably other forms of life), epigenetic 'marks' are considered to be a form of active, ongoing adaptation by the cell. Upregulation and downregulation, and expression and silencing of various genes in our DNA is the end result of chemical processes such as methylation and histone de-acetylation (histone modifications).[114]

It is now known that both our DNA and certain forms of RNA in our cells acquire epigenetic marks. Since viruses tend to be DNA, double-stranded RNA, single-stranded RNA, segmented RNA, or other forms of genetic sequences, they are very likely to be subject to their own versions of epigenetic modifications.

If you consider the entry of a virus into a cell as a new organism that has somewhat shared "goals" (the virocell), and since epigenetic marks are continually updating in our cells in response to our environment (e.g., food, liquid, and air intake), virions that enter into cells are likely to be subject to their own epigenetic modifications.

As a virus infects a cell, it must contend with that cell's defenses and the host's immune system, thus engaging in a power struggle for control of the cell's processes, and fending off competitor virions (other viruses), competitor bacteria, and other organisms. To successfully infect a cell

189

and produce viral progeny requires significant adaptation, and what is "learned" from that adaptation likely manifests itself in epigenetic marks, making the likelihood and methods of the viral progeny slightly or significantly changed by the infection event.

Changes in viral action and viral genetic sequences have been observed for many viruses, especially HIV.[115] People infected with HIV have been shown to produce many mutated forms of HIV after a period of time of infection.[115]

I hypothesize that viral infection happens from wild-type viruses (which are swarms of dozens, hundreds, or thousands of quasispecies), but also that successive infections, once inside a host, further diversify, shape, hone, and force adaptation of the infecting viral swarm, producing a whole new set of quasispecies to ensure successful infection in the future.

QUESTION 30:

"Would an isolate of an RNA virus that has only one sequence be less effective at establishing an infection than a wild-type virus that has thousands of different slight variants or quasispecies?"

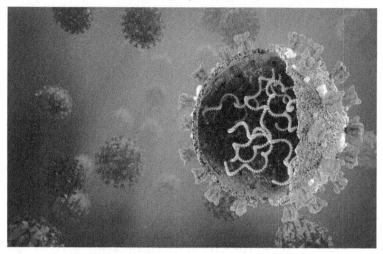

James Shapiro: (bio page 202)

On basic principles, I would say yes, because variability and diversity are generally far more effective biologically than uniqueness and homogeneity.[116]

Richard Allen White III: (bio page 209)

I think that natural selection would favor the ones that have more diversity. My hypothesis is that if you started with more diversity, then you'd probably end up with more; there could be recombination and single nucleotide polymorphisms (SNPs), and the result would probably be a more successful viral infection.

With influenza, there are multiple strains of the virus infecting the same cell, like in the case of pigs that have an avian virus in swine origin, or four different viruses infecting the same cell at some point. There could be a quadruple recombination event that allows for the creation of a new virus. This is what happened with swine-origin influenza in 2009.

Luis Villarreal: (bio page 203)

You wouldn't be able to make such a thing if you were to use the native polymerases because they just don't have the fidelity. As soon as you propagate it as a plaque, it starts to generate variability and diversity. The only way you could do that would be to clone it as a DNA, because the machinery of DNA maintains fidelity much better than RNA.

Then, you would have a uniform sequence that you could generate into RNA. The RNA that would be made off the DNA template would have much less variation in it than it would naturally have as an infection, or as a quasispecies in nature. As soon as you propagate it as a virus, it would generate variation.

With HIV and some of the other RNA viruses, treatment with an antiviral will typically lead to selection of variation of the sequence as it tries to adapt and replicate in the presence of that specific antiviral.[117] This means you pick up particular variations (or mutations, you would call them), and when you take that antiviral selection off, you propagate without that, and the past experience of that population is retained as a minority in the quasispecies.

The quasispecies remembers what it's been through and what selections have occurred by keeping them as minorities of the quasispecies.[118] For example, the live poliovirus vaccine that is used so effectively to contain poliovirus in humans, has within it the pathogenic version of the virus that is capable of causing disease. It is kept in check by the other members of the quasispecies that it can't outcompete. This vaccine is providing you with a glimpse of the fully pathogenic version of the virus that is being kept in check.

Richard Jacobs:

In addition to viral variants acting as memory, they act as a viral immune system, because some of the variants will be more resistant to cellular and host defenses.[119] This is why I hypothesize that being infected with a viral isolate of only one genetic sequence will not be as successful as being infected by a wild-type virus.

The same scenario likely applies to vaccines; the efficacy of vaccines can likely be increased by creating vaccines that have many genetic variants, creating a more successful defense against a diversified, wild-type virus infection.

CO-AUTHOR BIOS

Contributor: Eugene V. Koonin
Senior Investigator, National
Center for Biotech Information
www.ncbi.nlm.nih.gov
koonin@ncbi.nlm.nih.gov

Eugene Viktorovich Koonin is a Russian-American biologist and Senior Investigator at the National Center for Biotechnology Information (NCBI). He is a recognised expert in the field of evolutionary and computational biology.

Koonin gained a Master of Science in 1978 and a PhD in 1983 in molecular biology, both from the Department of Biology at Moscow State University. His PhD thesis, titled "Multienzyme organization of encephalomyocarditis virus replication complexes", was supervised by Vadim I. Agol.

From 1985-1991, Koonin worked as a research scientist in computational biology in the Institutes of Poliomyelitis and Microbiology at the USSR Academy of Medical Sciences, studying virus biochemistry and bacterial genetics. In 1991, Koonin moved to the NCBI, where he has held a Senior Investigator position since 1996.

Koonin's principal research goals include the comparative analysis of sequenced genomes and automatic methods for genome-scale annotation of gene functions.

He also researches in the application of comparative genomics for phylogenetic analysis, reconstruction of ancestral life forms and building large-scale evolutionary scenarios, as well as mathematical modeling of genome evolution.

Contributor: Forest Rohwer
Microbial Ecologist, Professor of
Biology, San Diego State University
https://coralandphage.org
frohwer@gmail.com

My research is on ecosystems, and the two ecosystems that I am most interested in are coral reefs and the human body.

Originally, it was on coral reefs because I was doing marine ecology and looking at viruses in seawater. I wanted something that stayed in the same place. Corals were a good mixture, because they are marine, but they are stuck to the bottom, so they can't move around much. They do turn out to be great analogs for the human body and particularly for mucosal surfaces.

Right now, I'm looking at a disease of the lungs called cystic fibrosis, which is caused by a genetic change in a human that creates a sticky mucus which gets colonized by viruses and bacteria, and eventually clogs up the lungs. I am looking at the ecology of that, coming up with ways of understanding what is happening in the disease, and looking for potential therapeutics based on information we get from genomic and metagenomic methods.

The other thing we are doing associated with humans is a sampling of different environmental reservoirs looking for SARS-CoV-2, the cause of

COVID. We are trying to get an idea of how common it is in the environment.

In terms of the coral reefs, I'm working on applying what we have learned in order to determine why coral reefs are degrading, and whether we can come up with restoration methods for coral reefs.

Contributor: Denis Noble

Evolutionary Biologist

www.denisnoble.com

denis.noble@dpag.ox.ac.uk

Denis Noble is a very well-spoken physiologist and evolutionary biologist, and has been involved in science for far a long time. He's worked on the mechanisms of the heart, and is a giant in the world of science.

Denis is professor emeritus and co-director of computational physiology at Oxford University. One of the pioneers of Systems Biology and developed the first viable mathematical model of the working heart in 1960. Over 600 articles in academic journals, including Nature, Science, PNAS, Journal of Physiology.

Denis Noble was Chairman of the IUPS (International Union of Physiological Sciences) World Congress in 1993, and Secretary-General of IUPS from 1993-2001. He was President of IUPS from 2009 to 2017. His previous publications include the seminal set of essays The Logic of Life (Boyd and Noble, OUP 1993), and he played a major role in launching the Physiome Project, one of the international components of the systems biology approach. Science magazine included him amongst its review authors for its issue devoted to the subject in 2002.

Education:

Noble was educated at Emanuel School and University College London (UCL). In 1958 he began his investigations into the mechanisms of heartbeat. This led to two seminal papers in Nature in 1960 giving the first proper simulation of the heart. From this work it became clear that there was not a single oscillator which controlled heartbeat, but rather this was an emergent property of the feedback loops in the various channels. In 1961 he obtained his PhD working under the supervision of Otto Hutter at UCL.

Research

Noble's research focuses on using computer models of biological organs and organ systems to interpret function from the molecular level to the whole organism. Together with international collaborators, his team has used supercomputers to create the first virtual organ, the virtual heart.

As secretary-general of the International Union of Physiological Sciences 1993–2001, he played a major role in launching the Physiome Project, an international project to use computer simulations to create the quantitative physiological models necessary to interpret the genome, and he was elected president of the IUPS at its world congress in Kyoto in 2009.

Noble is also a philosopher of biology, and his books The Music of Life and Dance to the Tune of Life challenge the foundations of current biological sciences, question the central dogma, its unidirectional view of information flow, and its imposition of a bottom-up methodology for research in the life sciences.

Contributor: Marilyn Roossinck
Professor, Plant Pathology Center
for Infectious Disease Dynamics
https://roossincklab.com
mjr25@psu.edu

I took a microbiology course and was introduced to the bacteriophage Lambda, which was the first virus I ever really knew anything about. I was so impressed by how clever it was, just this little package of genes in a protein coat. I decided to change my career and become a virologist.

I did my PhD in human virology, working on the Hepatitis B virus. I didn't really want to stay in animal or human research because I felt like plants offered an area where you could do a lot more work. For example, you can have a thousand genetically identical hosts to work with, and they are inexpensive and no one minds when you put them in the blender, either.

I decided to make a switch into plant virology, which I studied for a long time. There are really nice systems in plants that allow for experimental evolution. Then, I got involved in fungal viruses. Right now, my lab is only doing one project on fungal virus, and we are winding up all the other stuff.

Contributor: Paul Turner
Rachel Carson Professor of Ecology
and Evolutionary Biology at Yale
https://medicine.yale.edu/
paul.turner@yale.edu

When I was in graduate school in the early 1990s, I happened to be out in California, UC Irvine. At that time, the HIV epidemic was really taking off in LA and I decided to shift my focus more towards viruses and away from bacteria, at least a bit. That's what kind of set me on my path.

I began my graduate studies in Ecology and Evolutionary Biology at the University of California, working with Richard E. Lenski as my PhD advisor. When Lenski's research group transferred to Michigan State University in 1991, I moved with the group and completed my PhD in Zoology in 1995. In doing so, I became only the fifth African American to receive a PhD in evolutionary biology.

I'm interested in how viruses physically interact with cells and how those kinds of interactions can drive both virus evolution, as well as the ability for the cells and the host to resist virus attacks. What's different about the work that I do is those interests span not only the bacteriophages that infect bacteria, but also viruses that infect organisms (such as humans) and act as pathogens.

Contributor: James Shapiro

Department of Biochemistry and
Molecular Biology, University of Chicago
https://shapiro.bsd.uchicago.edu
jsha@midway.uchicago.edu

James A. Shapiro is Professor of Microbiology in The Department of Biochemistry and Molecular Biology at the University of Chicago. He received his B.A. in English Literature from Harvard in 1964 and his Ph.D. in Genetics from Cambridge University in 1968 under Prof. W. Hayes, FRS.

After postdoctoral fellowships at Institut Pasteur with Prof. Francois Jacob (1967-1968) and Harvard Medical School with Prof. Jonathan Beckwith (1968-1970), he served as Invited Professor in the School of Biological Sciences at the University of Havana, Cuba (1970-1972). At the University of Chicago since 1973, he was Darwin Prize Visiting Professor at the University of Edinburgh (1993).

In 2001, he received an O.B.E. from Queen Elizabeth for services to the Marshall Scholarship Program. He is a founding member of the web site, www.TheThirdWayofEvolution.com, intended to make the public aware of scientific alternatives to both Intelligent Design and Neo-Darwinism. He has published pioneering books on mobile genetic elements, natural genetic engineering, bacterial multicellularity, and read-write genome evolution.

Contributor: Luis Villarreal
Professor Emeritus, Molecular
Biology and Biochemistry
www.thethirdwayofevolution.com
lpvillar@uci.edu

Virology was something I backed into. I started my career without any deep objectives in terms of where I was going; I only knew that I was interested in science. When I first started coming out of East LA, I just wanted to get a job doing something scientific. I started training as a medical technologist at a junior college, but within a year, that got very dull and I transferred to Cal State University in Los Angeles.

I changed my major to biochemistry, which required me to redo a lot of the courses that I had already taken. This was a rather annoying aspect, but it seemed worthwhile at the time, so I did it and I was able to succeed.

Around the time that I was completing my training in biochemistry, I took a class in advanced biochemistry and read a book by Lehninger that had just come out. This book dealt with complicated protein structures and crystals, and contained images. One of the images my professor put up at the time was a crystalline array of an RNA virus in the cytoplasm of an infected cell.

At the time, I was very intrigued by the interface between life and chemistry. Seeing the image of that crystal, in a sense, crystallized in my mind that viruses were at that junction; they were the interface between life and chemistry. They were chemical entities, yet they had all these characteristics of life. This understanding started me on my way as an undergraduate. Eventually I would enter graduate school and do post doc studies in virology. I have been in that area ever since.

Contributor: St Patrick Reid
Assistant Professor,
University of Nebraska
www.unmc.edu
patrick.reid@unmc.edu

St. Patrick Reid: My research is primarily on what we call host-pathogen interactions. I am interested in looking at what host proteins a virus has to recruit in order to replicate and successfully reproduce after it has infected a cell.

I was born in Jamaica. My family moved to the USA when I was six years old. I grew up in Brooklyn, New York, and attended the University of Rochester for my undergrad, and Mt. Sinai School of Medicine for my PhD. My PhD studies were on Ebola virus. I also did a post-doctoral fellowship in France under Victor Volchkov, who was studying Ebola virus. I then returned to the USA and did another post-doctoral fellowship at the US Army Medical Research Institute of Infectious Disease (USAMRIID) at Fort Detrick, where I completed more work on Ebola.

Eventually, I shifted my focus towards chikungunya virus. In 2016, I became an assistant professor at the University of Nebraska Medical Center where I'm now working. Our lab focuses on chikungunya and Ebola, and now on SARS-CoV-2 as well.

Contributor: William B. Miller, Jr.
Evolutionary Biologist
www.themicrocosmwithin.com
wbmiller1@cox.net

I came from a family where it was actually understood that I was going to college, so the question was just about what I would choose to do. My father was an accountant and could never function well in groups. To some degree, I thought I might be the same. I was not crazy about the content of corporate America, but I loved science. Let's not forget; this was a long time ago, when physicians were independent practitioners and not in these mega practices like they are today.

Today's medical life bears little resemblance to medical life when I had to make a decision. It was a combination of doing something I loved and something I thought would give me personal freedom of action within the professional sphere. That is what got me started in medicine, and I loved it.

I had a terrific career, but I was always interested in other things along the way. In radiology, you are a common denominator for all of medical practice; all the specialties come to you for you to determine — based on the body part that they are interested in — what's going on. Unless you've deeply sub-specialized as a radiologist, you are in a fulcrum and get to see all kinds of medical things.

You have a medical vocabulary that is uncommonly broad compared to most physicians and even most general practitioners. This is because within the imaging sphere, you have to not only know the general things, but an awful lot of sub-specialty-specific things.

Here is the point: through imaging, I was beginning to see hard patterns of overlap, particularly in terms of how infectious disease could mimic cancer. I really couldn't understand why that was based on the medical evidence or scientific evidence that was available at the time.

I started to look up certain things about the way cells communicate, which was a very new topic back then. I was trying to understand why we had recurring disease patterns. It should all make sense if you had a very good theory.

Then I met a girl named Sue. I was at a medical meeting in Chicago, where I was attending eight or 10 hours of lectures per day. By the afternoon of the third day, I was exhausted. I couldn't think anymore, and I said to my partner, "Let's play hooky this afternoon and go to the art museum or the field museum, you choose."

He chose the field museum. When you walk into that major rotunda — that huge opening space of the field museum in Chicago — you will see Sue. Sue is probably the best articulated Tyrannosaurus rex in the world, and it knocked me out. I started to read the caption and couldn't quite understand how this particular species stayed the same for six million years.

As an expert in imaging, I really knew my bone anatomy; looking at that Tyrannosaurus rex skeleton for the first time, I noticed that the ribs were a lot like human ribs, and the same with the humerus, which is the long upper arm bone. Not only did it look a good deal like it, but even the ridges where the muscles insert were alike.

There was a huge difference in scale when it came to the pelvis, but even it was very similar to the human pelvis. I didn't understand the concept of conservation of features, and I raised some questions to my partner. I'm a very bright guy, but he just thought I was an idiot for asking. He thought all the questions get answered by knowing that sufficient random motion over a long period of time could yield the Tyrannosaurus rex. I found that absurd, even though it satisfied my great partner.

Contributor: Dr. Richard Allen White III

Founder, Raw Molecular Systems

www.rawmolecularsystems.com

raw937@gmail.com

Dr. Richard Allen White III has over 15 years of experience in next-generation sequencing, multi-omics, and computational biology. He's been here from the beginning to make sure everything's running smoothly.

Right now, he and his team are focusing on developing diagnostics and bio-therapies to combat COVID-19, as well as re-purposing bacteriophage or bacterial viruses to tackle multi-drug resistance.

He is currently a tenure-track Assistant Professor in the Department of Bioinfomatics and Genomics at University of North Carolina at Charlotte (UNCC). He is a start-up biotech founder of RAW Molecular Systems (RMS) INC out of Concord, North Carolina.

Prior to that he had worked with Janet and Christer Jansson on soil/rhizosphere multiomics at the PNNL. He briefly describes himself as a computational and molecular virologist trying to expand the "Systems," approach to ecosystems.

Contributor: Guenther Witzany
Biocommunication and
Natural Genome Editing
www.biocommunication.at
witzany@sbg.at

Guenther Witzany is an author of biocommunication, natural genome editing, and viruses as essential agents of life. He is a philosopher who specializes in the philosophy of science, the philosophy of language and communication, and the philosophy of biology.

In the late 1980s, he proposed the concept of life as a communicative structure, where cells and organisms organize and coordinate themselves through communication processes. He developed a new philosophy of biology based on the current empirical data in all domains of cellular life, virology and RNA-biology.

In a series of books, he applied his theory of biocommunication to all organismic kingdoms. Theory of biocommunication is the first and only theory that integrates all living agents empiristically in a non-reductionistic and non-mechanistic way and opens a ground breaking understanding of life through analyses of sign-mediated interactions within and between cells, tissues, organs, and organisms.

Contributor: Paul Offit
Director, Vaccine
Education Center
www.paul-offit.com
paul.offit@gmail.com

I started working in the 1980s. The first person who got me interested in viruses was Stanley Plotkin. In the year prior to my fellowship on infectious diseases at the children's hospital in Philadelphia, he had basically invented the RA 27/3 strain of the rubella vaccine. He was also interested in seeing whether he could do work on the rotavirus for the purposes of making a rotavirus vaccine. Dr. Plotkin was a leader in the field. The vaccine book is called Plotkin's Vaccines.

I was asked by the National Academy of Medicine to write a biography of Dr. Reece Hilleman, who in many ways, is the father of modern vaccines. I'm on the NIH active group that was put together by Francis Collins, and we are working on a piece for Science Translational Medicine about prospects for a safe COVID-19 vaccine.

I'm also on the FDA's vaccine advisory committee, and we are going to be meeting at the end of October to discuss whether or not one or more vaccines will get approval under emergency authorization. I've kept busy.

Contributor: Shiraz Shah

Senior Researcher, COPSAC

www.copsac.com

shiraz.shah@dbac.dk

I completed my master's degree in bioinformatics, which basically involves using computers to analyze protein and DNA sequences. Most living organisms on earth contain tremendous amounts of information within their DNA. A bacterium is typically three or four megabytes, and the human genome is about four gigabytes.

All of that is written in a biological code, so to speak, and as human beings, we have been learning to decode it since the first genomes started coming out in the early 90s. There is so much information in biology once you start sequencing DNA.

I began my PhD in 2007 and was employed at Copenhagen University to do some bioinformatics on a newly-discovered immune system in bacteria. Bacteria use this immune system to protect themselves against viral infections, because bacteria can get sick and die from viral infections.

In fact, I think every other bacterium dies every day from a viral infection. It's very lethal for a bacterium to get infected by a virus, so bacteria need to defend themselves against viruses. They have a number of different types of immune systems to do that; some of them fall under what we call innate immune mechanisms, and others fall under what we call adaptive immune mechanisms.

One of the adaptive immune responses that bacteria have are CRISPR-Cas systems. Back then, not a lot was known about CRISPR-Cas systems. We were sequencing different bacteria and archaea, and looking at the contents of the genome in terms of which genes were there.

We could see that we have the CRISPR systems on the genome because we could see that the bacteria were basically saving bits and pieces of sequences from the different viruses that had infected them over the past decade or so. The bacteria had encoded resistance against a series of different viruses, and one bacterium could have memory units against a hundred different viruses.

My assignment was to try to figure out—just by looking at the DNA sequence of the area where the CRISPR-Cas immune system is encoded—what we can determine about the interactions that the bacterium had against different viruses that it'd been infected by over the last decade, and the different types of CRISPR-Cas immune systems that exist in the different bacteria and archaea. CRISPR-Cas systems exist in many different versions depending on which bacterium or archaea you are looking at.

Back in 2008 or 2009, we were just starting to figure out what types of CRISPR-Cas immune systems exist. Today, one particular type—either Type 2A or Type 2C—is being used in human genome engineering. The entire diversity of CRISPR-Cas systems is a lot larger, and just one of them at the moment is being used as technology.

Contributor: Adolfo Garcia-Sastre
Director, Global Health and
Emerging Pathogens Institute
https://icahn.mssm.edu/
adolfo.garcia-sastre@mssm.edu

I became interested in viruses while going to college for biochemistry. In Spain, I completed my PhD thesis, which dealt with the enzymatic activity of a protein of a particular virus. I read more about viruses and became fascinated by how simple they are, and how they are able to reproduce themselves by using the machinery of a cell.

That's what made me decide to do my postdoctorate in virology. When I came to the states, I was accepted as a postdoc in a very good lab in Mount Sinai, and the rest is history. I think it was part chance and part fascination at the right time.

We are doing a lot of work on COVID-19 because that's pressing. In general, the questions that we are asking are not necessarily related to COVID-19, but more about how viruses are successful, how they are able to take over the host and cause disease, and whether there is anything we can do in order to prevent some of the viruses for which there are no therapeutics and no vaccine.

My work is also about gaining tools from the study of viruses and using them to better understand the biology of ourselves. It's a research program, mainly focused on virus biology, which is not so different

214

from general biology. Virus biology allows you to simplify some things, because viruses are so limited in the number of genes that they utilize.

This allows you to more easily ask questions about complex organisms that have many genes, like us. Studying viruses is a way to simplify things to look for answers in more complex organisms, while at the same time understand the virus' biology and see whether we can do something about preventing viral pathogens.

Contributor: Joseph R. Masci

Clinical Professor, Global Health

www.mountsinai.org

Joseph.Masci@mssm.edu

When I was in medical school, I was very inspired by the infectious diseases division and the people who worked in it. After I finished my residency in medicine, I decided to do a fellowship in infectious diseases, which was right at the beginning of the HIV/AIDS epidemic.

I came to Elmhurst Hospital in Queens, New York, where we were seeing many of these patients. The AIDS program developed in a lot of places over the next few years, and as time went by, other things also became very important.

After the 9/11 attack, there was concern about bioterrorism, so I focused on looking for strategies to detect and prevent bioterrorism. The only Ebola patient in New York was at Bellevue Hospital; I became very interested in Ebola and served as an advisor to the city hospital system.

The COVID-19 pandemic came as something of a surprise, but not a total surprise. It had a major impact in New York City, especially early on, and now it is all over the world. I'm very focused on planning for the response and looking at treatments and preventions for COVID-19.

Contributor: Jeremy Barr
Head, Bacteriophage Biology
Research Group
https://thebarrlab.org
jeremy.barr@monash.edu

I was really lucky in that I knew what I wanted to do from an early age. As a kid, I was always interested in science and actually fell in love with insects. I was obsessed with them from about age eight to 10, and read all sorts of books about them. During my time at university, I drifted into microbiology and eventually settled on bacteriophages.

Looking back, I would arguably say that bacteriophages are the insects of the microbial world. They're incredibly diverse, unique, extremely exciting, and they always come with different surprises. I've always been interested in the weird or quirky aspects of biology, and certainly that's what drew me to viruses.

In my lab at Monash University, we look at bacteriophages, which are viruses of microbes, specifically viruses of bacteria. Our lab looks at phages in this concept of tripartite symbiosis. Phages don't only prey on their bacteria, but also interact with the human host. The concept that viruses, bacteria, and the human host have all of these intricate symbioses is a focus of the lab. We are looking at phages in the context of bacteria, but also in the context of the human and mammalian host.

Contributor: Lily Wang
Instructor, Researcher
https://icahn.mssm.edu
lily.wang@mssm.edu

I have been in the virology field for about 14 years. I was doing work on the herpes virus in my PhD program, and afterwards, I came to New York at Mount Sinai School of Medicine to study HIV cell-to-cell spread. At present, I am still working on HIV in Dr. Benjamin's lab, and I also mentor some PhD students for their lab training and techniques.

Contributor: Nathalie Gontier
Director, Applied Evolutionary
Epistemology Lab
https://ciencias.ulisboa.pt/en
nlgontier@ciencias.ulisboa.pt

Nathalie Gontier is a researcher for the Faculty of Science of the University of Lisbon, sponsored by the Portuguese Foundation for Science and Technology under law 57/2016, and she is an Integrated Member of the Center for Philosophy of Sciences where, since 2012, she is the director of AppEEL – The Applied Evolutionary Epistemology Lab that was founded with the support of the John Templeton Foundation.

She has a background in both Philosophy and Comparative Science of Cultures (Cultural Anthropology), and she holds a PhD in Philosophy of Science. She is the Editor-in-Chief of the Springer Nature Book Series Interdisciplinary Evolution Research, Associate editor for Evolutionary Biology, Advisory Editorial Board Member for Theoria et Historia Scientiarum, and Review Editor for Frontiers in Psychology/Theoretical and Philosophical Psychology. She seats on the permanent board of the Protolang Conference Series, she is a Member of the Third Way of Evolution, and a collaborator for the Astra project.

Previously, she has been an Invited Associate Researcher for the Autonomous section for History and Philosophy of Science at the Faculty of Science of the University of Lisbon, a part-time Professor of

Philosophy of Science at the Dutch Free University of Brussels (Belgium); with the support of the Marie Curie Actions under Framework 7 of the European Union, she has worked at the Division of Paleontology of the American Museum of Natural History (New York City, USA); she has been a Post-Doctoral Fellow for the Portuguese Fund for Science and Technology; a Research Assistant for the Fund for Scientific Research Flanders (Belgium); and a Research Fellow at the Austrian Konrad Lorenz Institute.

Her publications include a monograph on theories on the origin and evolution of life that deviate from the standard paradigm (in Dutch), and she has edited several books and academic journals on topics of biological, linguistic and sociocultural evolution.

Contributor: Gareth Brady
Ussher Assistant Professor, Clinical
Medicine, Trinity College Dublin
https://www.tcd.ie
bradyg1@tcd.ie

My interest is in studying viruses that are very efficiently able to target innate immunity. All viruses develop some skills in doing this, but some are better than others. I did my PhD on vaccinia virus, which was the vaccine strain used to eradicate variola, the causing agent of smallpox. It became quite clear that it wasn't a great virus to study for the purposes of finding out how to turn off human pathways, because manifestly, this virus doesn't do it. I guess that's why it works as such a good vaccine strain; it drives the immune system nuts.

There is one virus called molluscum contagiosum that has specifically adapted to the human infection, which no other viruses are known to do. It only infects humans, and it's completely different from vaccinia virus. It was also very understudied before I started working on it. The most interesting thing about it is that it doesn't cause disease, which means it is very good at getting around the immune system.

Most virus disease is caused by the reaction against the virus by the immune system rather than what the virus is actually doing by itself. Unfortunately, this makes it a funding liability, because asking for funding to work on a virus that doesn't cause disease is an assailable thing.

My approach is to try to understand how it goes about evading the immune system and switching off the pathways that we use to detect and respond to a virus. Perhaps, this will reveal new things about the pathways. Viruses are evolving things, but are targeted specifically to the rate-limiting steps of these pathways.

They are like an investigative probing on the pathways to try to figure out the best ways to inhibit. If the objective is to find novel ways of switching off pro-inflammatory pathways, there's no better way to go about it than to study a virus that has evolved to specifically target the best bits.

Contributor: Michael Betts
Professor of Microbiology, Pennsylvania
Institute for Immunology
www.bettslab.org
betts@pennmedicine.upenn.edu

My interest is in understanding how humans combat pathogens with immune responses, and particularly, adaptive immune responses with T cells and B cells. We mostly work on T cells. Historically, we've focused on immunology in the context of an HIV infection.

In addition, we are now studying SARS-CoV-2 infection and working in areas of basic human immunology. We are trying to understand how T cells move around in the body and what happens to them when they go into different places.

This all pertains to how those cells will function in response to an invading viral pathogen or pathogenic challenge which takes place in potentially many different areas of the body.

Contributor: Nils G. Walter
Professor of Chemistry and
Biological Chemistry
https://sites.lsa.umich.edu/walter-lab/
nwalter@umich.edu

Nils Walter: I was trained in Germany as a chemist with emphasis on biochemistry. I received my undergrad degree, master's degree (at the time it was called a diploma), and PhD in Germany. For my PhD, I worked at the Max Planck Institute for Biophysical Chemistry. At the time, Nobel laureate Manfred Eigen was one of the directors, and I worked with him directly. He passed away in early 2019.

With Eigen, I worked on in vitro evolution which is the idea that you can have Darwinian selection, a process that would mimic how life works with replication, mutation, and selection from the diverse sequences you get through the mutation out of that replication, and that through cycles of Darwinian evolution, you can eventually evolve a new function in the test tube.

My attraction to this idea is why I joined his lab. I was the biochemist who was responsible for building an in vitro evolution machine that later went to the technical museum in Munich as a museum piece. From this experience, I learned that molecules, just as viruses, can actually evolve over time, but in a test tube. I also learned that there is a parallel between this in vitro evolution and what we now see has emerged as a virus,

SARS-CoV-2, and continues to mutate in infected humans. It is fascinating to see that connection.

After my PhD on in vitro evolution, I moved to the US as a post-doctoral fellow at the University of Vermont. I learned about RNA enzymes, or ribozymes for short. They are connected to in vitro evolution because they could have been the starting point for how life formed on earth. This is because RNA is the unique molecule across biology that has a feature of carrying genetic information, like it is in the genome of the COVID-19 agent, but it can also act like an enzyme as proteins are do.

RNA therefore kind of bridges DNA and proteins, and as such, it solves the chicken-and-egg problem. It allows us to imagine how life on earth could have started in an RNA World, where only RNA enzymes (i.e., "ribozymes") existed, replicated, mutated, and ultimately evolved over time. It's these ribozymes that I studied as a post-doc, and on which I started my own lab in 1999 at the University of Michigan, where I am still today. I am in chemistry, but I have multiple appointments in biophysics, biological chemistry, and across campus.

I co-founded the Center for RNA Biomedicine, which is about studying RNA with the hope of deriving medicines from it, such as RNA-based vaccines as seen in action against COVID-19, or as drugs targeted to cure diseases. The famous Spinal Muscular Atrophy disease can now be fought with a modified RNA molecule that fixes what we would call a splicing defect, and helps toddlers who have SMA to survive for years.

There are clearly opportunities to utilize our knowledge of RNA to cure diseases, which is the goal of the Center for RNA Biomedicine that I currently co-direct. We aim to bring together researchers from different disciplines to utilize more of the power of RNA for medicine down the road.

Contributor: Robert Siegel
Professor of Microbiology and
Immunology, Stanford University
https://profiles.stanford.edu/robert-siegel
http://web.stanford.edu/~siegelr/photo.html

Whenever we tell stories about our past, they are "just-so" stories that we make up so that our lives make sense. I do think one of the main things that got me interested in this field is the fact that the period when I was an undergraduate and graduate student was a very exciting time for molecular biology and virology.

The very first DNA sequences were coming out, and since researchers were trying to sequence small things, the first published sequence of any biological entity was for a virus. We were learning about viruses at a phenomenal pace and that certainly got me excited. At that time, I heard many lectures on viruses; I clearly remember lectures by Kathy Dana when I was in graduate school. She talked about a monkey virus (SV40) and her work in that area had a significant impact upon me.

Interestingly enough, I think my career sort of turned on a subsequent conversation that I had with an administrator. I had the opportunity to teach an undergraduate class in 1981. I wrote up two different class proposals: one was on molecular evolution and the other was on human virology. The administrator did not know a lot about science, but I put my proposals on her desk, and she said "viruses." I have been teaching my Humans and Viruses class ever since.

I have had a very remarkable career in that I have been able to focus primarily on teaching and advising. I have also done a number of research projects along the way. Working with several students, we recently published reviews of the biomedical literature regarding several aspects of the coronavirus, including a potential therapeutic that is an IL-6 inhibitor, and a meta-analysis of the research on asymptomatic and presymptomatic transmission. My most recent textbook chapter is on the classification of human viruses.

Most of my research has focused on how to be the most effective communicator I can be. I am always doing experiments when I teach and trying to learn more. This is more pertinent than ever when trying to teach in the Zoomosphere.

Contributor: Shervin Takyar

Associate Professor of Pulmonary,
Critical Care, Sleep Medicine, Yale
https://medicine.yale.edu
seyedtaghi.takyar@yale.edu

Syed Taqi Takyar, who goes by Shervin, is at Yale University and has a PhD in Microbiology and Molecular Biology from the University of Queensland, Australia. He has worked on projects that include viral vectors and has a lot of interests.

Shervin did his PhD in microbiology and molecular biology in The University of Queensland, Australia. During his PhD he worked and published on a variety of projects including developing a new lentiviral vector based on JDV (Jembrana Disease Virus), translational regulation in HCV by small RNA-binding molecules and the viral core protein, and RNA-protein interactions in positive strand RNA viruses.

During this time, he was also involved in cloning the Australian isolate of HCV with Dr Eric Gowans. His findings in these projects were published in a variety of journal including PNAS, Hepatology, and Journal of Molecular Biology. His next stop was a postdoctoral fellowship with Prof. Harry Noller at the RNA Center in UCSC where he delved deeper into the RNA world and studied the helicase activity of the ribosome during translation. His work was well received and published in Cell.

Contributor: Matthew Frieman
Associate Professor, University of
Maryland School of Medicine
www.umaryland.edu
matt.frieman@gmail.com

My research is on coronaviruses. We are taping this in the middle of a coronavirus pandemic and it just so happens that I've been working on coronaviruses since around 2004 when I started my post-doc at the University of North Carolina in Chapel Hill. I started on SARS-1 working on pathogenesis, host factors, and viral proteins that affect disease.

Then I started at University of Maryland School of Medicine in Baltimore. In 2009, I started my own lab still working on SARS. In 2012, we worked on MERS coronavirus. We are merging that work into SARS-CoV-2 and trying to study all aspects of therapeutic design, pathogenesis, how the virus works, and how the disease replicates.

We are really trying to understand a bit more about the virus than we already do.

Lindsay Hoeschen

Editor

Speakeasy Marketing, Inc.

lindsay@speakeasymarketinginc.com

Five years of undergraduate study in philosophy and the sciences, and four years as a freelance writer and editor has taught Lindsay Hoeschen at least one solid lesson: when the cat walks across the keyboard more than once, it's obligatory break time.

Since graduating from Portland State University School of Honors in 2014, Lindsay has pursued myriad interests along a nonlinear path, but always with a steadfast loyalty to the practice of critical thinking, self-reflection, and creativity.

References

1. Microbiology by numbers. *Nat Rev Microbiol* **9**, 628 (2011). https://doi.org/10.1038/nrmicro2644
2. Dolgin E. The secret social lives of viruses. Nature. 2019 Jun;570(7761):290-292. doi: 10.1038/d41586-019-01880-6. PMID: 31213694
3. Nomaguchi, M., Fujita, M., Miyazaki, Y., & Adachi, A. (2012). Viral tropism. *Frontiers in microbiology*, *3*, 281. https://doi.org/10.3389/fmicb.2012.00281
4. María Bergua, Sung-Hwan Kang, Svetlana Y. Folimonova (2016) Understanding superinfection exclusion by complex populations of Citrus tristeza virus. *Virology,* 499, 331-339. http://www.sciencedirect.com/science/article/pii/S0042682216302926
5. Patrick, F. (2012). Virocell Concept, The. *ELS.* doi:10.1002/9780470015902.a0023264
6. Nardacci, R., Perfettini, JL., Grieco, L. *et al.* Syncytial apoptosis signaling network induced by the HIV-1 envelope glycoprotein complex: an overview. *Cell Death Dis* **6**, e1846 (2015). https://doi.org/10.1038/cddis.2015.204
7. (n.d.). Retrieved January 15, 2021, from https://www.historyofinformation.com/detail.php?id=2620
8. Coffin JM, Hughes SH, Varmus HE, editors. Retroviruses. Cold Spring Harbor (NY): Cold Spring Harbor Laboratory Press; 1997. The Place of Retroviruses in Biology. Available from: https://www.ncbi.nlm.nih.gov/books/NBK19382/
9. Mougari, S., Sahmi-Bounsiar, D., Levasseur, A., Colson, P., & La Scola, B. (2019). Virophages of Giant Viruses: An Update at Eleven. *Viruses*, *11*(8), 733. https://doi.org/10.3390/v11080733'
10. Griffiths D. J. (2001). Endogenous retroviruses in the human genome sequence. *Genome biology*, *2*(6), REVIEWS1017. https://doi.org/10.1186/gb-2001-2-6-reviews1017
11. Gray M. W. (2012). Mitochondrial evolution. *Cold Spring Harbor perspectives in biology*, *4*(9), a011403. https://doi.org/10.1101/cshperspect.a011403
12. Lorenzo Subissi, Mory Keita, Samuel Mesfin, Giovanni Rezza, Boubacar Diallo, Steven Van Gucht, Emmanuel Onuche Musa, Zabulon Yoti, Sakoba Keita, Mamoudou Harouna Djingarey, Amadou Bailo Diallo, Ibrahima Soce Fall, Ebola Virus

Transmission Caused by Persistently Infected Survivors of the 2014–2016 Outbreak in West Africa, *The Journal of Infectious Diseases*, Volume 218, Issue suppl_5, 15 December 2018, Pages S287–S291, https://doi.org/10.1093/infdis/jiy280

13. Madelain, V., Baize, S., Jacquot, F., Reynard, S., Fizet, A., Barron, S., Solas, C., Lacarelle, B., Carbonnelle, C., Mentré, F., Raoul, H., de Lamballerie, X., & Guedj, J. (2018). Ebola viral dynamics in nonhuman primates provides insights into virus immuno-pathogenesis and antiviral strategies. *Nature communications*, *9*(1), 4013. https://doi.org/10.1038/s41467-018-06215-z

14. Esther Nolte-'t Hoen, Tom Cremer, Robert C. Gallo, and Leonid B. Margolis, Extracellular Vesicles and Viruses: Are They Close Relatives?, PNAS August 16, 2016 113 (33) 9155-9161; first published July 18, 2016; https://doi.org/10.1073/pnas.1605146113

15. Yong, E. (2020, August 10). Immunology Is Where Intuition Goes to Die. Retrieved January 22, 2021, from https://www.theatlantic.com/health/archive/2020/08/covid-19-immunity-is-the-pandemics-central-mystery/614956/

16. Nomoto A. (2007). Molecular aspects of poliovirus pathogenesis. *Proceedings of the Japan Academy. Series B, Physical and biological sciences*, *83*(8), 266–275. https://doi.org/10.2183/pjab.83.266

17. Janeway CA Jr, Travers P, Walport M, et al. Immunobiology: The Immune System in Health and Disease. 5th edition. New York: Garland Science; 2001. Pathogens have evolved various means of evading or subverting normal host defenses. Available from: https://www.ncbi.nlm.nih.gov/books/NBK27176/

18. Fleischmann WR Jr. Viral Genetics. In: Baron S, editor. Medical Microbiology. 4th edition. Galveston (TX): University of Texas Medical Branch at Galveston; 1996. Chapter 43. Available from: https://www.ncbi.nlm.nih.gov/books/NBK8439/

19. Taylor, B. S., Sobieszczyk, M. E., McCutchan, F. E., & Hammer, S. M. (2008). The challenge of HIV-1 subtype diversity. *The New England journal of medicine*, *358*(15), 1590–1602. https://doi.org/10.1056/NEJMra0706737

20. Mi S, Lee X, Li X, Veldman GM, Finnerty H, Racie L, LaVallie E, Tang XY, Edouard P, Howes S, et al.: Syncytin is a captive retroviral envelope protein involved in human placental morphogenesis. Nature 2000, 403:785-789

21. Schaumburg, C. S., Held, K. S., & Lane, T. E. (2008). Mouse hepatitis virus infection of the CNS: a model for defense, disease, and repair. *Frontiers in bioscience : a journal and virtual library*, *13*, 4393–4406. https://doi.org/10.2741/3012

22. Lavi E, Gilden DH, Highkin MK, Weiss SR. The organ tropism of mouse hepatitis virus A59 in mice is dependent on dose and route of inoculation. Lab Anim Sci. 1986 Apr;36(2):130-5. PMID: 3009966.

23. Lauring, A. S., & Andino, R. (2010). Quasispecies theory and the behavior of RNA viruses. *PLoS pathogens*, *6*(7), e1001005. https://doi.org/10.1371/journal.ppat.1001005

24. Pfeiffer, J., & Kirkegaard, K. (n.d.). Increased Fidelity Reduces Poliovirus Fitness and Virulence under Selective Pressure in Mice. Retrieved January 22, 2021, from https://www.ncbi.nlm.nih.gov/pmc/articles/PMC1250929/

25. University of British Columbia. (2019, March 28). The bigger the evolutionary jump, the more lethal cross-species diseases could be. *ScienceDaily*. Retrieved January 22, 2021 from www.sciencedaily.com/releases/2019/03/190328080357.htm

26. Wilen, C. B., Tilton, J. C., & Doms, R. W. (2012). HIV: cell binding and entry. *Cold Spring Harbor perspectives in medicine*, *2*(8), a006866. https://doi.org/10.1101/cshperspect.a006866

27. Callaway, E. (2020, September 08). The coronavirus is mutating - does it matter? Retrieved January 22, 2021, from https://www.nature.com/articles/d41586-020-02544-6

28. van Sluijs, L., Pijlman, G. P., & Kammenga, J. E. (2017). Why do Individuals Differ in Viral Susceptibility? A Story Told by Model Organisms. *Viruses*, *9*(10), 284. https://doi.org/10.3390/v9100284

29. Services, V. (n.d.). What determines susceptibility to virus infection? Retrieved January 22, 2021, from https://virologyresearchservices.com/2019/07/29/susceptibility-to-virus/

30. Wylie KM, Mihindukulasuriya KA, Zhou Y, Sodergren E, Storch GA, Weinstock GM. Metagenomic analysis of double-stranded DNA viruses in healthy adults. BioMed Central Biology, online Sept. 10, 2014.

31. *Scutti S (2014-11-20)*. "Why Some People Are Naturally Immune To HIV". *Medical Daily*. Retrieved 2021-01-22.

32. Moosavi, A., & Motevalizadeh Ardekani, A. (2016). Role of Epigenetics in Biology and Human Diseases. *Iranian biomedical journal*, *20*(5), 246–258. https://doi.org/10.22045/ibj.2016.01

33. Khanal, S., Ghimire, P., & Dhamoon, A. S. (2018). The Repertoire of Adenovirus in Human Disease: The Innocuous to the Deadly.

Biomedicines, 6(1), 30.
https://doi.org/10.3390/biomedicines6010030

34. Kansas State University. (n.d.). Retrieved January 22, 2021, from https://www.ksvdl.org/resources/news/diagnostic_insights/may201 6/cattle.html

35. Measles. (n.d.). Retrieved January 22, 2021, from https://www.pediatricweb.com/webpost/iframe/MedicalConditions _433.asp?tArticleId=172#8

36. Hamming, I., Timens, W., Bulthuis, M. L., Lely, A. T., Navis, G., & van Goor, H. (2004). Tissue distribution of ACE2 protein, the functional receptor for SARS coronavirus. A first step in understanding SARS pathogenesis. *The Journal of pathology, 203*(2), 631–637. https://doi.org/10.1002/path.1570

37. Aubry, F., Dabo, S., Manet, C., Filipović, I., Rose, N., Miot, E., . . . Lambrechts, L. (2020, November 20). Enhanced Zika virus susceptibility of globally invasive Aedes aegypti populations. Retrieved January 22, 2021, from https://science.sciencemag.org/content/370/6519/991

38. Lang, A. S., Zhaxybayeva, O., & Beatty, J. T. (2012). Gene transfer agents: phage-like elements of genetic exchange. *Nature reviews. Microbiology, 10*(7), 472–482. https://doi.org/10.1038/nrmicro2802

39. Lazzaro, B., & Clark, A. (n.d.). Rapid evolution of innate immune response genes. Retrieved January 22, 2021, from https://oxford.universitypressscholarship.com/view/10.1093/acprof :oso/9780199642274.001.0001/acprof-9780199642274-chapter-20

40. Melchjorsen, J., Matikainen, S., & Paludan, S. R. (2009). Activation and evasion of innate antiviral immunity by herpes simplex virus. *Viruses, 1*(3), 737–759. https://doi.org/10.3390/v1030737

41. Bacteriophage Makes *Vibrio Cholerae* Deadly , *Laboratory Medicine*, Volume 28, Issue 1, 1 January 1997, Pages 8–9, https://doi.org/10.1093/labmed/28.1.8

42. Aleksandrowicz, P., Marzi, A., Biedenkopf, N., Beimforde, N., Becker, S., Hoenen, T., Feldmann, H., & Schnittler, H. J. (2011). Ebola virus enters host cells by macropinocytosis and clathrin-mediated endocytosis. *The Journal of infectious diseases, 204 Suppl 3*(Suppl 3), S957–S967. https://doi.org/10.1093/infdis/jir326

43. Kramer, D. (n.d.). Discovering New Cell Types One at a Time. Retrieved January 23, 2021, from

https://www.nature.com/scitable/blog/bio2.0/discovering_new_cell_types_one/

44. Manning, A. J., & Kuehn, M. J. (2011). Contribution of bacterial outer membrane vesicles to innate bacterial defense. *BMC microbiology*, *11*, 258. https://doi.org/10.1186/1471-2180-11-258

45. Baluška, F., & Miller, W. B., Jr (2018). Senomic view of the cell: Senome *versus* Genome. *Communicative & integrative biology*, *11*(3), 1–9. https://doi.org/10.1080/19420889.2018.1489184

46. Sakai, T., Nishimura, S. I., Naito, T., & Saito, M. (2017). Influenza A virus hemagglutinin and neuraminidase act as novel motile machinery. *Scientific reports*, *7*, 45043. https://doi.org/10.1038/srep45043

47. *Pathak KB, Nagy PD (2009).*"Defective Interfering RNAs: Foes of Viruses and Friends of Virologists". *Viruses. 1 (3): 895–919. doi:10.3390/v1030895. PMC 3185524. PMID 21994575.*

48. *Kado, C.I. Origin and evolution of plasmids. Antonie Van Leeuwenhoek* **73**, 117–126 (1998). https://doi.org/10.1023/A:1000652513822

49. *Goldsby, H.J., Dornhaus, A., Kerr, B., & Ofria, C. (2012).* Task-switching costs promote the evolution of division of labor and shifts in individuality. *Proceedings of the National Academy of Sciences, 109*, 13686 - 13691.

50. *Sicard, A., Michalakis, Y., Gutiérrez, S., & Blanc, S. (2016).* The Strange Lifestyle of Multipartite Viruses. *PLoS pathogens*, *12*(11), e1005819. https://doi.org/10.1371/journal.ppat.1005819

51. *Dearborn, A. D., Laurinmaki, P., Chandramouli, P., Rodenburg, C. M., Wang, S., Butcher, S. J., & Dokland, T. (2012).* Structure and size determination of bacteriophage P2 and P4 procapsids: function of size responsiveness mutations. *Journal of structural biology*, *178*(3), 215–224. https://doi.org/10.1016/j.jsb.2012.04.002

52. Erez, Z., Steinberger-Levy, I., Shamir, M. *et al.* Communication between viruses guides lysis–lysogeny decisions. *Nature* **541**, 488–493 (2017). https://doi.org/10.1038/nature21049

53. Buchrieser J, Dufloo J, Hubert M, Monel B, Planas D, Rajah MM, Planchais C, Porrot F, Guivel-Benhassine F, Van der Werf S, Casartelli N, Mouquet H, Bruel T, Schwartz O. Syncytia formation by SARS-CoV-2-infected cells. EMBO J. 2020 Dec 1;39(23):e106267. doi: 10.15252/embj.2020106267. Epub 2020 Nov 4. PMID: 33051876; PMCID: PMC7646020.

54. Justin E. Silpe, Bonnie L. Bassler, A Host-Produced Quorum-Sensing Autoinducer Controls a Phage Lysis-Lysogeny Decision,

Cell, Volume 176, Issues 1–2, 2019, Pages 268-280.e13, ISSN 0092-8674, https://doi.org/10.1016/j.cell.2018.10.059. (http://www.sciencedirect.com/science/article/pii/S0092867418314582)

55. Memory gene goes viral. (2018, January 16). Retrieved January 25, 2021, from https://www.nih.gov/news-events/news-releases/memory-gene-goes-viral

56. Eigen, Manfred; Schuster, Peter (1979). "The Hypercycle". *Naturwissenschaften*. **65**(1): 7–41. doi:10.1007/bf00420631. ISSN 0028-1042.

57. *Eigen M (October 1971). "Selforganization of matter and the evolution of biological macromolecules". Die Naturwissenschaften. 58 (10): 465–523. Bibcode:1971NW.....58..465E. doi:10.1007/bf00623322. PMID 4942363.*

58. Ma, W., Kahn, R. E., & Richt, J. A. (2008). The pig as a mixing vessel for influenza viruses: Human and veterinary implications. *Journal of molecular and genetic medicine : an international journal of biomedical research*, *3*(1), 158–166.

59. Conniff, R. (2019, November 12). Discover Interview: E.O. Wilson. Retrieved January 27, 2021, from https://www.discovermagazine.com/planet-earth/discover-interview-eo-wilson

60. Parkin J, Cohen B (June 2001). "An overview of the immune system." *Lancet*. 357 (9270): 1777-89. doi: 10.1016/S0140-6736(00)04904-7. PMID 11403834. S2CID 165986

61. Page, M. (2019, March 30). Giant viruses have weaponised CRISPR against their bacterial hosts. Retrieved January 27, 2021, from https://www.newscientist.com/article/2197422-giant-viruses-have-weaponised-crispr-against-their-bacterial-hosts/

62. Muturi, E., Buckner, E., & Bara, J. (2017, February 24). Superinfection interference between dengue-2 and dengue-4 viruses in Aedes aegypti mosquitoes. Retrieved January 27, 2021, from https://onlinelibrary.wiley.com/doi/full/10.1111/tmi.12846

63. Folimonova S. Y. (2012). Superinfection exclusion is an active virus-controlled function that requires a specific viral protein. *Journal of virology*, *86*(10), 5554–5561. https://doi.org/10.1128/JVI.00310-12

64. Thaker, S.K., Ch'ng, J. & Christofk, H.R. Viral hijacking of cellular metabolism. *BMC Biol* **17**, 59 (2019). https://doi.org/10.1186/s12915-019-0678-9

65. Barr, J. (2017, August 30). A bacteriophages journey through the human body. Retrieved January 27, 2021, from https://onlinelibrary.wiley.com/doi/10.1111/imr.12565

66. Barr JJ, Auro R, Furlan M, Whiteson KL, Erb ML, Pogliano J, Stotland A, Wolkowicz R, Cutting AS, Doran KS, Salamon P, Youle M, Rohwer F. Bacteriophage adhering to mucus provide a non-host-derived immunity. Proc Natl Acad Sci U S A. 2013 Jun 25;110(26):10771-6. doi: 10.1073/pnas.1305923110. Epub 2013 May 20. PMID: 23690590; PMCID: PMC3696810.

67. Jon CohenMar. 16, 2., Jon CohenJan. 26, 2., Kai KupferschmidtJan. 26, 2., Lizzie WadeJan. 26, 2., Jon CohenJan. 25, 2., Jon CohenJan. 22, 2., . . . Sofia MoutinhoJan. 5, 2. (2017, December 10). Examining His Own Body, Stanford Geneticist Stops Diabetes in Its Tracks. Retrieved January 27, 2021, from https://www.sciencemag.org/news/2012/03/examining-his-own-body-stanford-geneticist-stops-diabetes-its-tracks

68. McAllister F, Khan MAW, Helmink B, Wargo JA. The Tumor Microbiome in Pancreatic Cancer: Bacteria and Beyond. Cancer Cell. 2019 Dec 9;36(6):577-579. doi: 10.1016/j.ccell.2019.11.004. PMID: 31951558.

69. Aizpurua-Olaizola O, Sastre Torano J, Pukin A, Fu O, Boons GJ, de Jong GJ, Pieters RJ. Affinity capillary electrophoresis for the assessment of binding affinity of carbohydrate-based cholera toxin inhibitors. Electrophoresis. 2018 Jan;39(2):344-347. doi: 10.1002/elps.201700207. Epub 2017 Oct 4. PMID: 28905402.

70. Secor, P. R., Burgener, E. B., Kinnersley, M., Jennings, L. K., Roman-Cruz, V., Popescu, M., Van Belleghem, J. D., Haddock, N., Copeland, C., Michaels, L. A., de Vries, C. R., Chen, Q., Pourtois, J., Wheeler, T. J., Milla, C. E., & Bollyky, P. L. (2020). Pf Bacteriophage and Their Impact on Pseudomonas Virulence, Mammalian Immunity, and Chronic Infections. *Frontiers in immunology, 11*, 244. https://doi.org/10.3389/fimmu.2020.00244

71. Locke, S. (2014, June 27). How viruses stay one step ahead of our efforts to kill them. Retrieved January 27, 2021, from https://www.vox.com/2014/6/27/5846900/how-viruses-stay-one-step-ahead-of-our-efforts-to-kill-them

72. Arnold, C. (2016, September 28). The Viruses That Made Us Human. Retrieved January 27, 2021, from https://www.pbs.org/wgbh/nova/article/endogenous-retroviruses/

73. Abedon S. T. (2019). Look Who's Talking: T-Even Phage Lysis Inhibition, the Granddaddy of Virus-Virus Intercellular Communication Research. *Viruses, 11*(10), 951. https://doi.org/10.3390/v11100951

74. OpenStax. (n.d.). Microbiology. Retrieved January 28, 2021, from https://courses.lumenlearning.com/microbiology/chapter/isolation-culture-and-identification-of-viruses/

75. Smith GP. Filamentous fusion phage: novel expression vectors that display cloned antigens on the virion surface. Science. 1985 Jun 14;228(4705):1315-7. doi: 10.1126/science.4001944. PMID: 4001944.
76. Badri T, Gandhi GR. Molluscum Contagiosum. [Updated 2020 Aug 8]. In: StatPearls [Internet]. Treasure Island (FL): StatPearls Publishing; 2020 Jan-. Available from: https://www.ncbi.nlm.nih.gov/books/NBK441898/
77. Pita JS, de Miranda JR, Schneider WL, Roossinck MJ. Environment determines fidelity for an RNA virus replicase. J Virol. 2007 Sep;81(17):9072-7. doi: 10.1128/JVI.00587-07. Epub 2007 Jun 6. PMID: 17553888; PMCID: PMC1951419.
78. Ayala-Nunez, N., Hoornweg, T., van de Pol, D. *et al.* How antibodies alter the cell entry pathway of dengue virus particles in macrophages. *Sci Rep* **6,** 28768 (2016). https://doi.org/10.1038/srep28768
79. Lant, K. (2018). C difficile Infections Increasing, Becoming More Difficult to Treat. Retrieved January 28, 2021, from https://www.hcplive.com/view/c-difficile-infections-increasing-becoming-more-difficult-to-treat
80. Forterre P. (2010). Defining life: the virus viewpoint. *Origins of life and evolution of the biosphere : the journal of the International Society for the Study of the Origin of Life*, 40(2), 151–160. https://doi.org/10.1007/s11084-010-9194-1
81. Broecker, F., & Moelling, K. (2019, January 14). Evolution of Immune Systems From Viruses and Transposable Elements. Retrieved January 28, 2021, from https://www.frontiersin.org/articles/10.3389/fmicb.2019.00051/full
82. Ni, G., Ma, Z., & Damania, B. (2018). cGAS and STING: At the intersection of DNA and RNA virus-sensing networks. *PLoS pathogens*, *14*(8), e1007148. https://doi.org/10.1371/journal.ppat.1007148
83. Watson RO, Bell SL, MacDuff DA, Kimmey JM, Diner EJ, Olivas J, Vance RE, Stallings CL, Virgin HW, Cox JS. The Cytosolic Sensor cGAS Detects Mycobacterium tuberculosis DNA to Induce Type I Interferons and Activate Autophagy. Cell Host Microbe. 2015 Jun 10;17(6):811-819. doi: 10.1016/j.chom.2015.05.004. Epub 2015 Jun 2. PMID: 26048136; PMCID: PMC4466081.
84. Mizutani Y, Kanbe A, Ito H, Seishima M. Activation of STING signaling accelerates skin wound healing. J Dermatol Sci. 2020 Jan;97(1):21-29. doi: 10.1016/j.jdermsci.2019.11.008. Epub 2019 Nov 26. PMID: 31813660.
85. Faruque, S. M., & Mekalanos, J. J. (2012). Phage-bacterial interactions in the evolution of toxigenic Vibrio

cholerae. *Virulence*, *3*(7), 556–565.
https://doi.org/10.4161/viru.22351

86. Harris, H. (2020, April 23). In defense of viruses: Most are harmless, and many can be beneficial to us. Retrieved January 28, 2021, from https://phys.org/news/2020-04-defense-viruses-harmless-beneficial.html

87. Cadwell K. (2015). The virome in host health and disease. *Immunity*, *42*(5), 805–813. https://doi.org/10.1016/j.immuni.2015.05.003

88. RNA metabolism. (n.d.). Retrieved January 28, 2021, from https://www.nature.com/subjects/rna-metabolism

89. Sender, R., Fuchs, S., & Milo, R. (2016). Revised Estimates for the Number of Human and Bacteria Cells in the Body. *PLoS biology*, *14*(8), e1002533. https://doi.org/10.1371/journal.pbio.1002533

90. Linares, D. M., Ross, P., & Stanton, C. (2016). Beneficial Microbes: The pharmacy in the gut. *Bioengineered*, *7*(1), 11–20. https://doi.org/10.1080/21655979.2015.1126015

91. Alam, M.T., Amos, G.C.A., Murphy, A.R.J. *et al.* Microbial imbalance in inflammatory bowel disease patients at different taxonomic levels. *Gut Pathog* **12,** 1 (2020). https://doi.org/10.1186/s13099-019-0341-6

92. Azzouz LL, Sharma S. Physiology, Large Intestine. [Updated 2020 Jul 27]. In: StatPearls [Internet]. Treasure Island (FL): StatPearls Publishing; 2020 Jan-.Available from: https://www.ncbi.nlm.nih.gov/books/NBK507857/

93. Rafii, F., Sutherland, J. B., & Cerniglia, C. E. (2008). Effects of treatment with antimicrobial agents on the human colonic microflora. *Therapeutics and clinical risk management*, *4*(6), 1343–1358. https://doi.org/10.2147/tcrm.s4328

94. Oeggerli, P. (2019, December 17). How trillions of microbes affect every stage of our life-from birth to old age. Retrieved January 28, 2021, from https://www.nationalgeographic.com/magazine/2020/01/how-trillions-of-microbes-affect-every-stage-of-our-life-from-birth-to-old-age-feature/

95. Rezac, S., Kok, C. R., Heermann, M., & Hutkins, R. (2018). Fermented Foods as a Dietary Source of Live Organisms. *Frontiers in microbiology*, *9*, 1785. https://doi.org/10.3389/fmicb.2018.01785

96. Knezevic, J., Starchl, C., Tmava Berisha, A., & Amrein, K. (2020). Thyroid-Gut-Axis: How Does the Microbiota Influence Thyroid Function?. *Nutrients*, *12*(6), 1769. https://doi.org/10.3390/nu12061769

97. Microbes Help Produce Serotonin in Gut. (n.d.). Retrieved January 28, 2021, from https://www.caltech.edu/about/news/microbes-help-produce-serotonin-gut-46495

98. Dietzgen, R. G., Mann, K. S., & Johnson, K. N. (2016). Plant Virus-Insect Vector Interactions: Current and Potential Future Research Directions. *Viruses, 8*(11), 303. https://doi.org/10.3390/v8110303

99. Barr, J. J., Auro, R., Furlan, M., Whiteson, K. L., Erb, M. L., Pogliano, J., Stotland, A., Wolkowicz, R., Cutting, A. S., Doran, K. S., Salamon, P., Youle, M., & Rohwer, F. (2013). Bacteriophage adhering to mucus provide a non-host-derived immunity. *Proceedings of the National Academy of Sciences of the United States of America, 110*(26), 10771–10776. https://doi.org/10.1073/pnas.1305923110

100. Lin G, Lee B, Haggarty BS, Doms RW, Hoxie JA. CD4-independent use of Rhesus CCR5 by human immunodeficiency virus Type 2 implicates an electrostatic interaction between the CCR5 N terminus and the gp120 C4 domain. J Virol. 2001 Nov;75(22):10766-78. doi: 10.1128/JVI.75.22.10766-10778.2001. PMID: 11602718; PMCID: PMC114658.

101. Anekal, S. G., Zhu, Y., Graham, M. D., & Yin, J. (2009). Dynamics of virus spread in the presence of fluid flow. *Integrative biology : quantitative biosciences from nano to macro, 1*(11-12), 664–671. https://doi.org/10.1039/b908197f

102. Li F. (2016). Structure, Function, and Evolution of Coronavirus Spike Proteins. *Annual review of virology, 3*(1), 237–261. https://doi.org/10.1146/annurev-virology-110615-042301

103. Cronin, L. (2019). Working out our messy differences at the origin of life. Retrieved January 29, 2021, from https://chemistrycommunity.nature.com/posts/44980-working-out-our-messy-differences-at-the-origin-of-life

104. Harish, A., Abroi, A., Gough, J., & Kurland, C. (2016). Did Viruses Evolve As a Distinct Supergroup from Common Ancestors of Cells? *Genome biology and evolution, 8*(8), 2474–2481. https://doi.org/10.1093/gbe/evw175

105. Higgs, P., Lehman, N. The RNA World: molecular cooperation at the origins of life. *Nat Rev Genet* **16,** 7–17 (2015). https://doi.org/10.1038/nrg3841

106. Nikaido, M., Noguchi, H., Nishihara, H., Toyoda, A., Suzuki, Y., Kajitani, R., Suzuki, H., Okuno, M., Aibara, M., Ngatunga, B. P., Mzighani, S. I., Kalombo, H. W., Masengi, K. W., Tuda, J., Nogami, S., Maeda, R., Iwata, M., Abe, Y., Fujimura, K., Okabe, M., … Okada, N. (2013). Coelacanth genomes reveal signatures

for evolutionary transition from water to land. *Genome research*, *23*(10), 1740–1748. https://doi.org/10.1101/gr.158105.113

107. What are viral vectors? (n.d.). Retrieved January 29, 2021, from https://www.beckman.com/support/faq/research/what-are-viral-vectors

108. Brandes, N., & Linial, M. (2019). Giant Viruses-Big Surprises. *Viruses*, *11*(5), 404. https://doi.org/10.3390/v11050404

109. Zhang Wang, Martin Wu, Comparative Genomic Analysis of *Acanthamoeba* Endosymbionts Highlights the Role of Amoebae as a "Melting Pot" Shaping the *Rickettsiales* Evolution, *Genome Biology and Evolution*, Volume 9, Issue 11, November 2017, Pages 3214–3224, https://doi.org/10.1093/gbe/evx246

110. Yin, Y., & Fischer, D. (2008). Identification and investigation of ORFans in the viral world. *BMC genomics*, *9*, 24. https://doi.org/10.1186/1471-2164-9-24

111. Darwin's rabbit is revealing how the animals became immune to myxomatosis. (2019). Retrieved January 29, 2021, from https://www.nhm.ac.uk/discover/news/2019/february/darwins-rabbit-is-revealing-how-the-animals-became-immune-to-myxomatosis.html

112. Viruses revealed to be a major driver of human evolution. (2016, July 13). Retrieved January 29, 2021, from https://www.sciencedaily.com/releases/2016/07/160713100911.htm

113. Löber U, Hobbs M, Dayaram A, Tsangaras K, Jones K, Alquezar-Planas DE, Ishida Y, Meers J, Mayer J, Quedenau C, Chen W, Johnson RN, Timms P, Young PR, Roca AL, Greenwood AD. Degradation and remobilization of endogenous retroviruses by recombination during the earliest stages of a germ-line invasion. Proc Natl Acad Sci U S A. 2018 Aug 21;115(34):8609-8614. doi: 10.1073/pnas.1807598115. Epub 2018 Aug 6. PMID: 30082403; PMCID: PMC6112702.

114. Weinhold B. (2006). Epigenetics: the science of change. *Environmental health perspectives*, *114*(3), A160–A167. https://doi.org/10.1289/ehp.114-a160

115. Tough, R. H., & McLaren, P. J. (2019). Interaction of the Host and Viral Genome and Their Influence on HIV Disease. *Frontiers in genetics*, *9*, 720. https://doi.org/10.3389/fgene.2018.00720

116. Reed, D., & Frankham, R. (2003). Correlation between Fitness and Genetic Diversity. *Conservation Biology, 17*(1), 230-237. Retrieved January 29, 2021, from http://www.jstor.org/stable/3095289

117. van Opijnen T, de Ronde A, Boerlijst MC, Berkhout B. Adaptation of HIV-1 depends on the host-cell environment. PLoS One. 2007 Mar 7;2(3):e271. doi: 10.1371/journal.pone.0000271. PMID: 17342205; PMCID: PMC1803020.

118. Ruiz-Jarabo, C. M., Arias, A., Baranowski, E., Escarmís, C., & Domingo, E. (2000). Memory in viral quasispecies. *Journal of virology*, *74*(8), 3543–3547. https://doi.org/10.1128/jvi.74.8.3543-3547.2000

119. Domingo, E., Sheldon, J., & Perales, C. (2012). Viral quasispecies evolution. *Microbiology and molecular biology reviews : MMBR*, *76*(2), 159–216. https://doi.org/10.1128/MMBR.05023-11

BONUS CONTENT

To listen to the all of the full audio interviews and transcripts that comprise this book (and more), visit: FindingGeniusFoundation.org and select "Publications" from the menu.

Permissions and Rights

All photos have been sourced and paid for from www.shutterstock.com.

Ordering Information:
Quantity sales. Special discounts are available on quantity purchases by corporations, associations, and others. For details, contact the publisher at the address above.

Orders by U.S. trade bookstores and wholesalers. Please call (888) 988-7381 or visit www.findinggeniusfoundation.org.

Printed in the United States of America

Published in 2021

ISBN: 978-1-954506-03-9

Made in the USA
Coppell, TX
26 March 2021

52423670R10138

December 2008

Dear Family,

I am excited to share this Arzoian family history book with you. There are pictures of our family that will hopefully evoke memories.

In the summer of 1940, Tommy was a toddler and I, an 8 year old. Dad borrowed Uncle Marty's car and drove our family, plus grandma and grandpa to the East Coast. We went to New York and stayed at Jennie Garabedian's home. Jennie's father was Grandma Arzoian's brother (our children's great grandmother). During this trip, we visited the World's Fair in New York City, Niagara Falls, Boston, Connecticut and New Hampshire.

In New Hampshire, we stayed at grandpa's sister's farm. Peggy Kayajan's family and other families were there, too. On the way home, we stayed with relatives in Chicago. We had a grand picnic with our extended family. I did not know that the picnic was the primary reason for the trip. All of us were squished in the car. I sat on the floor and I lay above the back seat by the rear window. Tommy and I would trade seats between dad and grandpa. During our trip, we would stay at motels to get a good night's rest. Some nights, the trains passing by the motel would awake us.

Nazaret Arzoian (grandpa), Hazavart Rose Garabedian Arzoian (grandma), Toros Tom Arzoian (dad), Marian Mary Garo Garoian Arzoian (mom), Thomas Michael Tommy Tom and I enjoyed several weeks of togetherness, traveling through the United States and meeting many relatives.

Just a little history note about our grandparents, grandpa came to the U.S. first. Then he sent for grandma, Uncle Marty and dad from Garmery, Kharpert, Asia Minor. Dad was born in 1901 and he arrived in the U.S. when he was 7 years old.

Here are some notes about the pictures of our family:
Page 55 – left to right: 3rd adult is dad, and then grandpa with a cigar and Tommy was being held. The others were relatives. Mom (Mary Garo Arzoian) was not in the picture.
Page 56 – left to right: Peggy, me, Tommy & Cousin Helen. Back row, left to right: Peggy's Aunt Depajian, Margaret Kayajan (Peggy's fraternal grandmother), Grandma Arzoian, and Jennie.
Page 126 – Cousin Jennie is on the right.

If I gather more information, I will be sure to keep it together and share with you.

Love,

Jean

Jean Jeannie Elissa Geraldine Arzoian Kalem